Patrick Moore's Practical Astronomy Series

Springer
London
Berlin
Heidelberg
New York
Barcelona
Hong Kong
Milan
Paris
Singapore
Tokyo

Other titles in this series

The Observational Amateur Astronomer
Patrick Moore (Ed.)

Telescopes and Techniques
C.R. Kitchin

The Art and Science of CCD Astronomy
David Ratledge (Ed.)

The Observer's Year
Patrick Moore

Seeing Stars
Chris Kitchin and Robert W. Forrest

Photo-guide to the Constellations
Chris Kitchin

The Sun in Eclipse
Michael Maunder and Patrick Moore

Software and Data for Practical Astronomers
David Ratledge

Amateur Telescope Making
Stephen F. Tonkin

Observing Meteors, Comets, Supernovae and other
Transient Phenomena
Neil Bone

Astronomical Equipment for Amateurs
Martin Mobberley

Transit: When Planets Cross the Sun
Michael Maunder and Patrick Moore

Practical Astrophotography
Jeffrey R. Charles

AstroFAQs
Stephen F. Tonkin

Deep-Sky Observing
Steven R. Coe

The Deep-Sky Observer's Year
Grant Privett and Paul Parsons

Observing the Moon

Peter T. Wlasuk

With 86 Figures

Springer

Peter T. Wlasuk, FRAS
Department of Physics and Astronomy, Florida International University,
UP Campus, Miami, Florida 33185, USA

Dedicated to Patrick Moore, who has inspired the author
and countless others to observe and study the Moon

Patrick Moore's Practical Astronomy Series ISSN 1431-9756

ISBN 1-85233-193-3 Springer-Verlag London Berlin Heidelberg

British Library Cataloguing in Publication Data
Wlasuk, Peter
    Observing the moon. – (Practical astronomy)
    1. Moon – Observers' manuals
    I. Title
    523.3
ISBN 1852331933

Library of Congress Cataloging-in-Publication Data
Wlasuk, Peter, 1962–
    Observing the moon / Peter T. Wlasuk.
        p. cm – (Practical astronomy, ISSN 1431-9756)
    ISBN 1-85233-193-3 (alk. paper)
    1. Moon–Observers' manuals. I. Title. II. Series.
    QB581.W63 2000
    523.3–dc21   00-033820

Typeset by EXPO Holdings, Malaysia
Printed and bound at the Cromwell Press, Trowbridge, Wiltshire
58/3830–543210 Printed on acid-free paper SPIN 10739180

Preface

Why write another guide to observing the Moon? That was the question I was pondering as I began this project, having a fine collection of "classic" lunar guidebooks dating back to 1791 in my own library. As a Fellow of the Royal Astronomical Society (RAS), member of the American Astronomical Society's Division for Planetary Sciences (AAS DPS), and member of the American Geophysical Union (AGU), I am fortunate to know many professional lunar scientists who keep me up to date with developments in lunar science – contrary to public perception, lunar science has definitely not stagnated since the last Apollo, No. 17, left the surface of the Moon in December, 1972. I am also lucky to know many amateur lunar observers, who, like me, enjoy actually looking at the Moon with telescopes and imaging it with a wide variety of devices ranging from regular 35 mm cameras to video recorders and CCD cameras. My friends who study the Moon, whether in their professions or just for fun, gave me several reasons for doing "another" lunar guidebook.

First, the last lunar observer's guide of any length was published over ten years ago, and many reviewers noted that it was badly out of date even then. It contained little discussion of modern lunar geology, much less what we learned from the exploration of the Moon by spacecraft, or from the Apollo astronauts and the many scientists who studied the lunar rocks and soil they brought back to Earth. Two shorter works have appeared in recent years, both written by amateur astronomers, that, much to their credit, finally begin to discuss these exciting discoveries, but one of these is privately published and not widely available. I am not surprised that this is the case, for neither NASA nor lunar scientists went out of their way to explain breakthroughs in our understanding of the Moon's geology or origin. Even the late Carl Sagan, whom I greatly admired for his grasp of how important it is for scientists to excite the tax-paying public about the worthy work they do, called the Moon "boring"!

As a result, very few lay persons, even those who are interested in astronomy, can tell me what we have learned about the Moon since World War II, when the impact theory finally overtook the theory of lunar volcanoes, or "lunavoes" as the late Axel Firsoff colorfully called them, to explain the formation of the majority of lunar craters. I suspect that if Apollo were in full swing today, the public would be better informed, both because the science community is now far more skilled at public relations than it was in the early 1970s, and because of the Internet and other technological wonders that allow the public to access and share information that was unavailable only a short time ago.

A second reason for writing an updated guide to observing the Moon is really a whole set of reasons. Since the last guidebooks were published, the evolution of the personal computer (PC) and the emergence of video and CCD imaging technologies have brought to the amateur new techniques not even dreamt of back in the early 1970s, when I started to observe the Moon for myself. Compare the CCD images of the Moon shown in Chapter 12 with photographs of the same lunar surface features taken by the Lunar Orbiter spacecraft, long considered the "ultimate" lunar images, and you will see just how far amateur lunar observing and imaging have advanced. Presenting this excellent work, which so richly deserves an audience, is another good reason for writing a new lunar guidebook, and I hope that seeing what can be done will inspire others to take up lunar observing with a new enthusiasm. That is why I chose to include a lunar atlas (on the CD-ROM) produced by an amateur lunar imager, rather than just republishing NASA spacecraft photos that bear little resemblance to what the user of this handbook is likely to see in their own telescope.

A third reason for writing a new guide to the Moon is to correct the huge number of errors that plague the classic guidebooks. When I first started observing the Moon over twenty-five years ago, I thought there was something seriously wrong with my eyes, or perhaps it was my telescope – I simply could not see many of the details that the authors of those classic guidebooks told me I was supposed to see! In time, as my observing skills and confidence improved, I realized that the fault was not with my eyes or my telescope – the classic guidebooks, in their descriptions of lunar surface features, contain literally hundreds of errors. When I was employed at the Yale University Department of Astronomy in the 1980s, I had the privilege of working under the tutelage of Dorrit Hoffleit on the *Yale Bright Star Catalogue* and its *Supplement*, as well as the *Yale Catalogue of Trigonometric Parallaxes*.

From this cataloging work I learned how to find and correct errors, and resolved to apply what I had learned to correcting errors in the descriptions of lunar surface features. Professional lunar scientists may think it silly to worry about such things, but it is not – amateur lunar observers have long enjoyed knowing what to "look for" on the face of the Moon, and they are entitled to an accurate guidebook. For the past quarter-century I have observed the Moon through a wide variety of instruments, ranging from a fine 60 mm Japanese refractor made in the late 1950s to the giant 40-inch (1-meter) refractor at the University of Chicago's Yerkes Observatory in Williams Bay, Wisconsin. My principal instruments for observing the Moon today include a classic Celestron 10 *f*/13.5 Schmidt–Cassegrain from 1969, the year that Neil Armstrong took those first famous steps on the lunar soil, and two antique reflectors made by the English pioneer telescope-makers George Calver and George With.

The heart of this book is comprised by Chapters 4 to 7, which contain detailed descriptions of lunar surface features. As I observed each of these features over the years, I would be careful not to study any photographs of those features from spacecraft, lest I got some preconceived idea of what I was "supposed" to see. I did refer to the classic lunar guidebooks available to me, but with a critical eye – I learned not to assume that they were correct. After making careful notes and drawings of the features, I would next consult Lunar Orbiter or other high-resolution images of the features, to verify what I had observed, finally comparing both the photographic image and my description to the descriptions in the classic guidebooks.

I discovered that the authors of the classic guidebooks often copied their descriptions from one another, which resulted in the propagation of innumerable mistakes. If you read the scientific literature from the time these guidebooks were written, it is clear that there was an intense contest among the classic observers to outdo one another to "discover" and to be the first to report subtle "new" details of lunar topography. This competition led, I believe, to observers pushing their instruments and their eyes beyond what could be reliably observed or recorded. Occasionally, strange beliefs in lunar vegetation or active lunar volcanism undoubtedly influenced what observers saw, or thought they saw. It was once quite fashionable to report dark radial bands spiraling across many lunar craters, as well as vivid but ephemeral colors in lunar features that we now know to be inconsistent with the make-up of lunar rocks and soil as discovered by examining them at first hand in our geologic laboratories.

A fourth objective in writing this book combines two of the others – I wanted to relate what we have learned about lunar geology in the past fifty years to the descriptions of lunar surface features, so that amateur lunar observers can get more out of their observing sessions. As majestic as the Moon is through a telescope, it can quickly get boring to look at one crater after another with no understanding of how they formed, how they differ from other craters or other lunar surface features, or whether a particular crater (or other formation) raises any questions that baffle even today's professional lunar scientists.

The amateur lunar observer who understands at least a modicum of lunar geology and geologic history will not get bored with the Moon, ever, as there are always new mysteries to be unraveled. Today, many amateur lunar observers are discussing sophisticated problems of lunar geology, asking well-phrased questions about the Moon's topography and how it got that way. Many are contributing valuable data to lunar studies in areas such as lunar occultations.

There are many more reasons for a book like this one, but rather than list them all, I would like to finish by acknowledging just a few of the many persons who contributed to this work, by providing text or illustrations, or helpful advice: Sally Beaumont; Julius Benton; Bill Davis; Bill Dembowski; Maurizio Di Sciullo; Eric Douglass; Colin Ebdon; Harry Jamieson; Anthony Kinder, Librarian, British Astronomical Association; Joe LaVigne; Robert Levitt; Mary A. Noel, Lunar and Planetary Institute (LPI), Houston, Texas; Michael Oates; Bernard Wahl; John Westfall; Ewen Whitaker, Lunar and Planetary Laboratory, University of Arizona; Ted Wolfe; and the Lunar Sections of the Association of Lunar and Planetary Observers and of the British Astronomical Association. Unless indicated otherwise, these persons are all amateur lunar observers, and without their fine work this book would not have been possible. If I have forgotten to thank anyone, which is very possible given the many persons with whom I discuss the Moon, I apologize. I also wish to thank my wife Patty, for helping with the proofreading and illustrations, and for her encouragement. I hope you enjoy this book!

Peter T. Wlasuk, FRAS

Contents

Chapter 1

Introduction to the Moon

So you want to observe the Moon, or at least learn more about it? Let's assume at least one of those things is true, or you would not have purchased this book. Before we get to topics like lunar formations and geology, detailed descriptions of the most rewarding lunar features for observers like yourself, or more advanced topics such as developing your observing skills or imaging the Moon with camera, video camera or CCD detectors, it's a good idea to take a little time to learn some basic facts about our nearest celestial neighbor, especially its movements in space, as understanding them will allow you to get more out of your observing sessions farther down the road.

Vital Statistics

When it came to lunar studies, many nineteenth-century astronomy textbooks covered only topics like the size and shape and orbital mechanics of the Moon, often in excruciating detail. I shall avoid going to these extremes. After all, though astronomers from Sir Isaac Newton in the late seventeenth century to Ernest W. Brown and Dirk Brouwer in the twentieth century expended much intellectual effort on constructing accurate mathematical models to describe the complex motions of the Moon, the gory mathematical details of these models are of little interest to amateur astronomers – they are appreciated more by mathematicians and by astronomers who study the field known as celestial mechanics. Luckily, we can acquire a basic understanding of the Moon's dimensions and motions without having to take a PhD in math!

Many books on the Moon quote its diameter – 3476 km (2160 miles) – but don't tell us what's significant about this piece of information. Oddly, this figure tells us that the Moon is both big and small, depending how you look at it. Compared to other planet–satellite groupings in our Solar System, the Moon is much nearer the Earth (diameter 12 756 km, 7926 miles) in size than other moons are to their parent planets. So in this sense the Moon is "big." Planetary scientists think of the Earth–Moon system as a double planet, which may seem a strange idea at first. It certainly would have seemed strange to ancient astronomers who, before Galileo discovered the true nature of the Moon by observing its craters and mountains, thought of Luna as a tiny but perfect celestial orb having nothing in common with our own planet.

But the fact is, though the Moon is a little over one-fourth the diameter of the Earth and contains less than one-eighteenth of its mass (the Moon "weighs" 7.35×10^{22} kg, compared with the Earth's 6×10^{24} kg), the two bodies are still close enough in size and mass and close enough together to exert powerful influences over each other. For example, every schoolchild learns that the Moon's "pull" causes tides in our oceans; if the Moon were a lot smaller or farther way, the tides would be negligible. And the Earth's appreciable mass and proximity to the Moon keeps the latter body in a tight, roughly circular orbit, locked in a "captured rotation" that forces the Moon to always keep the same hemisphere facing our planet.

But, like everything else in our universe, this celestial dance, though fairly stable and unchanging for the time being, is not permanent – the Moon is slowly spiraling away from the Earth, at the rate of 3.2 cm ($1\frac{1}{4}$ inches) per year. This is another consequence of tidal friction between the two bodies, and the fact that the Earth rotates more quickly than the Moon orbits us, which together create a net force that tends to accelerate the Moon in the direction of its orbit. This process transfers angular momentum between rotation and revolution, so that, as the Moon slowly drifts away from us, the length of the Earth day is slowly increasing: the Earth's rotation is slowing down by – 0.002 s per century.

Another reason to think of the Earth and Moon as a double planet is that the Moon probably formed when a large impacting body, most likely an asteroid, struck the Earth, scooping up with it pieces of the Earth's crust and throwing the whole agglomeration into orbit around Earth, where it eventually coalesced into a new, tinier composite planet. So the Moon is probably part Earth, part something else. We shall talk more about this theory of the formation of the Moon in Chapter 3, "A Crash Course in Lunar Geology."

But the Moon's diameter – only about two-thirds of the distance from the east coast of the United States to the west coast – makes it a small body in absolute terms. Why is this important to the lunar observer? For at least two reasons I can think of: for putting lunar topography into proper perspective, and to allow us to imagine what it would be like to actually visit the Moon – how things would look to us if we were as lucky as the Apollo astronauts, and could explore the Moon up close and personal. If you flip through the detailed descriptions of lunar features in Chapters 4 through 7 of this book, you will see that a typical lunar crater has walls that rise some 3000 m (10 000 ft) above its floor. That's a pretty respectable altitude for a mountain here on Earth. But in relative terms it's much more impressive than that, since the Moon is only one-fourth the diameter of the Earth. In other words, that 3000 m high crater rim or lunar mountain would have an equivalent height of 12 000 m (40 000 ft) on our planet! By way of comparison, Mt Everest is "only" 8848 m (29 028 ft) high.

The great relative height of lunar mountains above the "flat" terrain should give you new respect for the tremendous energies and forces of the impacts that shaped so much of the Moon's topography. Unlike Earth mountains, which are built either by the gradual sliding of one tectonic plate over another or by volcanic eruptions, the Moon's mountains result from the impact of asteroids. Lunar mountain ranges, which are really just the rims of very large craters (even the lunar maria, or "seas", the dark, more or less circular areas that are so huge they are easily visible to the naked eye, are just large craters filled in with lavas), and the so-called central peaks often found inside craters are both consequences of impact events. Figure 1.1 shows the Apennines, a mountain range at the rim of the Mare Imbrium impact basin.

There is no evidence that plate tectonics played a major role in building lunar mountains; and lunar volcanism, when it was occurring very long ago, bore little resemblance to the active volcanoes, like the volcano on the Caribbean island of Montserrat, which we see erupting today on our planet. Lunar volcanism consisted of very gradual, very large-scale and comparatively gentle "sheet flows" of lavas that covered, rather than created, lunar mountains and hills. Again, because the Moon is smaller than the Earth, lunar craters take up a disproportionately large area compared with similar terrestrial features, and the lunar lava flows also appear more impressive.

But the Moon's tiny diameter in absolute terms has another interesting consequence – the smaller a spherical body is, the more violently curved its horizons are. In other words, for any observer the same height above the ground, the horizon on the Moon appears much closer than the Earth horizon we are familiar with. The closeness of lunar horizons profoundly affects the appearance of lunar features – if you are standing on the Moon itself. For example, if you were an astronaut standing inside a typical large lunar crater you might just assume that you would see towering crater walls all around you. But if the crater were large enough, you would see no walls at all – they would be over the horizon. You wouldn't even know you were in a crater unless your map told you so. In fact, once they were actually walking across the lunar soil, the Apollo astronauts sometimes had difficulty finding and identifying features on the Moon's surface for this very reason. From ground level, the features they were looking for rarely looked like they did in photographs taken from Earth of from spacecraft orbiting the Moon.

The Moon's extreme curvature also explains the so-called foreshortening effect, a concept every observer must understand and keep in mind when looking at features anywhere near the lunar limb. The closer a feature is to any part of the lunar limb – north, south, east or west – the more that feature will appear distorted. Features near the east or west limb are squashed in the east–west direction (from side to side), appearing much more elliptical than they really are. Features close to the north or south pole of the Moon are squashed in the north–south direction (from top to bottom). Figure 1.2 is an image of the Moon's extreme southern limb, illustrating the foreshortening effect.

Craters and maria that appear to be elliptical because they are near a limb are in reality almost always very circular. Foreshortening also affects the apparent distances between lunar features near the limbs – they appear more crowded together than

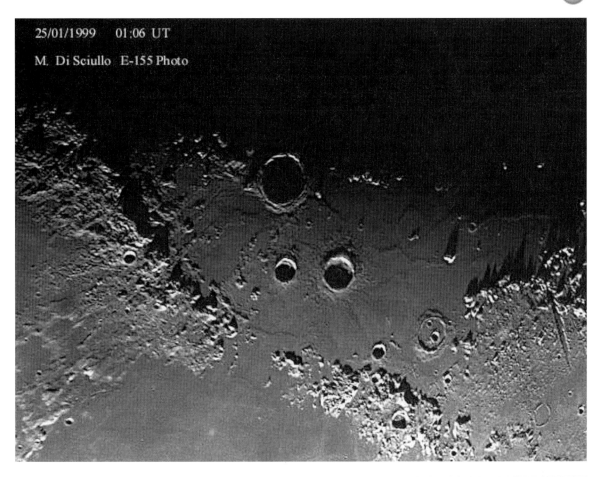

25/01/1999   01:06 UT

M. Di Sciullo E-155 Photo

Figure 1.1. The Apennines mountain range bordering Mare Imbrium. *CCD image by Maurizio Di Sciullo.*

they really are. This effect is especially obvious in the heavily cratered highlands near the Moon's south pole, where there seems to be an unusually high concentration of walled plains and smaller craters, an effect that is exaggerated by the absence of any mare regions to break up the chaotically cratered landscape.

The foreshortening effect is a geometric fact of life, an optical illusion, and indeed the Lunar Orbiter spacecraft that visited the Moon in the mid-1960s and extensively photographed its features from a vantage point directly above them confirmed that many craters that look elliptical from Earth are really quite round. Interestingly, before the Lunar Orbiters, Earth-bound astronomers had constructed special Moon maps that compensated for the fore-shortening effect. The lunar observer who consults any of the classical lunar observing guidebooks must remember that the dimensions quoted for features near the limbs are apparent measures only, and must be corrected for the foreshortening effect.

Enough said about the size of the Moon. What of its shape? This may sound rather obvious, but the Moon is round – *very* round. The Earth is often described as an oblate spheroid, meaning that, like many a middle-aged person, it bulges at the equator. The Moon shows no such appreciable equatorial bulge. Why the difference? The Earth's interior is largely in a fluid state, and this fluid builds up a sufficient centrifugal force as the Earth rapidly rotates on its axis to bulge out the relatively thin, solid crust of our planet. The Moon lacks this extensive fluid interior – it is basically a big chunk of rock – so it doesn't bulge much. It is, however, a little lopsided, depressed between 2 and 4 km on its Earth-facing side (called the *nearside*), and bulging away from the Earth 1 to 5 km on the *farside*. These figures include an overall tidal bulge of 1 km (0.6 miles) toward the Earth, caused by gravitational forces. Even when all of these measurements are taken into account, it may safely be said that the Moon is quite spherical.

Figure 1.2. The Moon's extreme southern limb, illustrating the foreshortening effect. *CCD image by Maurizio Di Sciullo.*

We've already seen that topographic features like mountains and crater rims have a higher profile on the Moon, relatively speaking, than on the Earth, so the Moon appears "rougher" than our planet. We also have to keep in mind that since the Moon has no appreciable atmosphere (just a low concentration of atoms and ions escaping from the surface) nor any "weather" to speak of, the forces that have reshaped the Earth's topography over the eons are largely absent on the Moon, so that we see its features much as they were billions of years ago.

The Moon's Motions in Space

The motions of the Moon fall into two main categories: its orbit about the Earth (and here we must also take into account the Moon's revolution about the Sun) and its so-called librations, or rockings back and forth, up and down. These motions explain a variety of phenomena of great interest to the lunar observer, from the Moon's phases, which tell us which features we will be able to observe on any given night, to lunar and solar eclipses.

Earlier I said that the Moon follows a "roughly circular" orbit about the Earth, and for purposes of envisioning phenomena like the Moon's phases this is a sufficient description. But of course the Moon must obey Kepler's laws of planetary motion, so its true orbital shape is that of an ellipse, and sometimes it is closer to Earth, sometimes farther away. At its closest approach, called *perigee*, the Moon is 363 263 km (225 727 miles) from us, while at its most distant retreat, called *apogee*, its separation from Earth increases to 405 547 km (251 995 miles). The apparent size of the Moon naturally varies with its approach or retreat from a minimum diameter of 29′ 26″ at apogee, to a maximum of 33′ 30″ at perigee (30′ or "30 minutes" is one-half a degree).

Roughly speaking, the Moon appears to subtend an angle of half a degree in the sky – also about the same angle subtended by the Sun, which is why we have solar eclipses. Every eighteen months or so on average, the Earth–Moon–Sun geometry is such that, along a narrow path somewhere on our planet, the new moon as seen from the Earth appears to intercept the Sun and perfectly cover it – this is a total solar eclipse. At other times, the Moon's apparent size is not quite big enough to completely cover the Sun's disk, then we have what's called an annular eclipse, the Sun's light forming a thin, bright ring around the black disk of the Moon.

But remember that the Moon is slowly moving farther and farther away from the Earth, as a result of tidal friction, so that someday, many centuries from now, the Moon will be so far away, even at perigee, that its apparent size will be insufficient to completely eclipse the Sun. In other words, we are lucky to be living at a time when the apparent size of the Moon as viewed from our planet is just right – not too big, and not too small – to barely cover the Sun's disk and allow us to see beautiful and scientifically interesting phenomena like solar prominences, huge loops of hydrogen gas, as well as the famous corona, that rarefied, hot outer halo of the Sun's atmosphere that glows at a temperature of 1.1 million °C.

By far the most obvious consequence of the Moon's motion around Earth is the phenomenon of lunar phases. Simply put, the Moon's phase is the fraction of the hemisphere facing us that is lit by the Sun, as seen by observers on the nighttime side of the Earth, the side that faces away from the Sun. The terminology is a little confusing because it is inconsistent. *New moon* is when the Moon is directly between the Earth and the Sun, so that all the sunlight falls on the farside of the Moon, which we can't see, while the Earth-facing side is dark. *Full moon* is when the Moon is directly opposite the Sun, with Earth perfectly sandwiched in between the two. But at full moon only half the Moon is illuminated – the whole Earth-facing half; the farside is completely dark at this phase. Between new and full moon comes *first quarter*, and *last quarter* is between full moon and new moon. These phases are correctly named, for in each case one-fourth of the lunar sphere is illuminated, and we see one half of the Earth-facing side brightly lit by the Sun's rays. First and last quarter occur when an imaginary line connecting the Moon with the Earth makes a 90° angle with a similar line connecting the Earth with the Sun. Figure 1.3 illustrates the phases of the Moon, from a beautiful old map.

Since the Moon takes 27.33 days to revolve once about the Earth, a period known as the *sidereal*

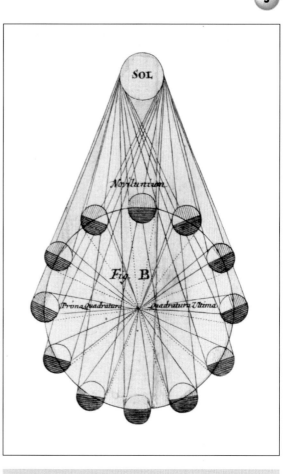

Figure 1.3. Diagram showing how the geometry of the Sun–Earth–Moon system illuminates different fractions of the Moon's Earth-facing hemisphere, which are known as the Moon's phases. From a 1742 copperplate engraving map appearing in an atlas compiled by the German astronomer Johann Gabriel Doppelmayer (1671–1750), published by the Homann heirs, a family responsible for producing many fine early terrestrial and celestial maps. *Author's collection.*

month, we often relate the Moon's phases to its *age* in days. For example, first quarter occurs when the Moon is said to be 6–7 days old, and full moon at about 14 days old, while last quarter is around 20–21 days. A new moon is zero days old. Any good astronomical almanac will tell you exactly how old the Moon is on any given date. As the Moon goes from new to full, we say its phase is *waxing*, meaning its phase is on the increase. From full to new, the phase is *waning* – decreases. Figure 1.4 shows the Moon a couple days before first quarter.

Anyone who has spent any time looking at the Moon realizes that it always keeps the same hemisphere turned Earthward. This is a rather unfortunate fact of celestial mechanics, known as *captured* the bulge would have to migrate over the lunar surface spinning beneath it!

This unhappy state of affairs would create what planetary geologists call a continuous plastic deformation of the viscous Moon as it struggled to spin against the restraining action of the tidal bulge. This would exert a braking force on the rapidly spinning Moon, just as a bicycle brake slows the wheel by pressing against the rotating wheel rim. The braking effect would be amplified as the young Moon cooled down, hardening its rocky outer layers, making it more and more difficult to rotate against the stubborn tidal bulge. Finally, over many eons, the Moon would naturally adopt a rotation rate equal to its orbital period, as the Moon would no longer have to rotate against the force of the tidal bulge if its rotation and orbit were synchronized. It is difficult to believe that today a tiny 1 km (0.6 mile) deformation in the lunar crust is the only clue to the dynamics of the Earth–Moon system.

From Fig. 1.3, it is apparent that the full moon rises just as the Sun sets – it will always be opposite the Sun in the sky. First quarter is "halfway between," so it rises roughly six hours before sunset, reaching the meridian right at sunset, setting roughly six hours after the Sun. The Moon at last quarter doesn't rise until around six hours after the Sun has set, and doesn't reach the meridian until sun-up. The new moon never appears to rise at all – it is approximately in the line of sight of the Sun itself, which is how we get solar eclipses. Because of the inclination of the Moon's orbit, the varying distance of the Moon from Earth and other factors, we don't get a total solar eclipse every time there is a new moon, and when conditions are right for an eclipse, only those on a narrow strip of the Earth's surface are treated to the spectacle.

Between new moon and first quarter, when the sunlit portion of the Moon is not terribly bright, you should look for the phenomenon known as *earthshine*. This is simply the gentle light reflected by the Earth onto the "dark" portion of the lunar disk, that is, the part not directly illuminated by the Sun itself. Earthshine is still sunlight, but sunlight that has already struck the Earth and been reflected at a lesser strength towards our satellite. It makes for a very beautiful scene, best observed with the naked eye or a pair of very low-power binoculars. Some lunar observers enjoy seeing how many features they can pick out just by earthshine alone – it's a fun challenge, and you should try it sometime.

The Moon's equator is inclined to its orbital plane by 6° 41′. Its orbital plane does not quite coincide with the plane of the Earth's orbit (the ecliptic), but is inclined by about 5°. This, combined with the

Figure 1.4. The three-day old Moon. *Photograph by Bob Levitt.*

rotation, which deprives us of enjoying the many wonders on the farside, the hemisphere permanently turned away from our planet. But what exactly is captured rotation? Remember that the Moon has a modest tidal bulge near the center of its nearside hemisphere. This bulge in the solid lunar rock is more stressful to the Moon in terms of energy and drag than the reverse effect of the Moon as observed in our ocean tides.

Let's suppose that shortly after the Earth–Moon system formed, some 4.5 billion years ago, the Moon's period of rotation was much greater than its orbital period. As the early Moon, which was probably hot and somewhat viscous, spun rapidly on its axis, the Earth-facing tidal bulge would, under the forces of gravitation, always have to point in the same direction relative to our planet. To do this, the Moon would have to rotate underneath this bulge, which would stay put. To put it another way,

Sun's gravitational pull, gives rise to a so-called precession of the Moon's orbit, meaning that the nodes (the points where the Moon's orbit and the Earth's intersect) gradually shift their position westward, making a complete revolution about the ecliptic every 18 years, 7 months or so. This tells us something about the Moon's travels through our skies.

The Moon's ascending node is where it crosses the ecliptic from south to north; its descending node is where is crosses the other way, north to south. If the Moon's orbit were in the same plane as the Earth's, it would vary its height above the southern horizon by 47° yearly, because the Earth's equator is inclined by 23°.5 to its orbital plane (±23°.5 is a total variation of 47°). But the Moon has an additional inclination of 5°, and when this is taken into account we see that it can attain a northerly or southerly declination of ±28°.5, making a total variation of 57°. What all this means is that the height of the Moon above the horizon varies more dramatically than that of the Sun.

You may well ask whether our knowledge of the Moon's motions allows us to determine the most favorable times for observing our natural satellite. To answer this question, we must first ask what we mean by "favorable." Three factors are involved here. The first two are really variants of the same phenomenon. First, the Moon should not be too close to the Sun, so observing a very young or very old Moon will be difficult no matter when we try to do it. I find it very challenging to get good steady images of lunar features for Moons younger than 3 days or older than 25 days. This is true, not because the Sun's light interferes with the lunar image, but because at these times the Moon is so close to the setting or rising Sun that it is very low in the western (evening) or eastern (morning) sky, forcing the observer to look through thick layers of our own atmosphere which badly degrade the Moon's image. In other words, when the Moon is very young or very old it never gets very high above the horizon while it is dark, and the features visible during these phases tend to be neglected by amateur observers.

This leads us to the second factor to consider, the Moon's closeness to the meridian. The meridian is the imaginary line (actually a "great circle") that runs from the north celestial pole, through the zenith (the apex of the celestial vault, the point directly over our heads) downward to the celestial south pole. When any heavenly body, be it a star, planet or the Moon, intersects the meridian, it is as "high in the sky" as it will get that evening, and thus most favorably placed for observation. The higher an object is in the sky, the fewer distorting effects the atmosphere will have on it. The Moon will cross the meridian at a time halfway between its rising and setting times, so if you know those times, it is easy to reckon when the Moon will be most favorably placed. Some astronomical almanacs will specifically tell you when the Moon will be on the meridian.

The third factor is the Moon's altitude above the southern (for northern-hemisphere observers) horizon. I used to live in upstate New York, at a latitude of about 43°N, from where the Moon and planets, which in general have orbits that follow the ecliptic, would often appear to skim the treetops to my south. They would rise in the southeast and set in the southwest, because I was well north of the equator. When I moved to Florida the first thing I noticed was how high overhead the Moon and planets suddenly were – my new latitude, 26°N, was closer to the ecliptic (and hence to the equator), so Solar System objects appeared to rise at a point closer to true east and set closer to true west. No longer did they skim the treetops to my south, but were always high enough above the southern horizon to be easily observed. Florida is an excellent place from which to observe the Moon and planets, not only because of its favorable latitude, but also because it is surrounded by water and so enjoys steady skies – a quality known as seeing that we will take up in Chapter 8.

So in general you will find it easier to observe the Moon if you live closer to the equator than if you live at higher latitudes, but of course few people are going to uproot themselves and move closer to the equator just so they can observe the Moon! But wherever you live, there are certain times of the year when the Moon will be higher above your southern horizon than others, depending on the steepness of the ecliptic at sunset or sunrise. Remember our brief discussion of the Moon's variation in declination, owing to the inclination of the Earth's equator to its orbit, and the inclination of the Moon's equator to its orbit? Taking into account these factors, we are able to predict when the Moon will be highest above the southern horizon during different phases.

For example, at first quarter the Moon is most favorably seen around the third week of March, the time of the spring equinox. Why is this so? Around this time, the Sun sets due west for the northern-hemisphere observer, or very close to it. From our discussion of Moon phases, we know that at first quarter the Moon will be on the meridian at this time. It will be high in the sky, in approximately the same spot the Sun will occupy three months later at the summer solstice. The northern section of the ecliptic is above the horizon at sunset, and when the first quarter Moon sets, it will do so in the north-west. From this time onward, the Moon between first quarter and full will gradually decrease in altitude as it crosses the meridian, and continues to do so until the autumnal equinox, when the full moon rises due east at sunset, and sets due west. The Moon is right on the celestial equator at this point.

What is the situation at the time of the summer solstice? The Sun is now as far above the celestial equator as it will get, so it sets in the northwest. The first quarter Moon that crossed the meridian as the solstice Sun set is now on the celestial equator – it's at "medium" height above the southern horizon, and will rise and set due east and west. The full moon spends much less time moving across the sky during the summer solstice, for it is at the point in the sky where the Sun is found during winter solstice, so it rises in the southeast and sets in the southwest.

Going through these motions, we can see that the best time to observe the 3–4 day old (waxing) crescent is late April/early May for northern-hemisphere observers. The full moon attains its greatest altitude above the horizon at the time of the winter solstice, while at last quarter the Moon is best placed during the third week of September, at the time of the autumnal equinox. Finally, a 25–26 day old (waning) crescent Moon will be most favorably placed for northern-hemisphere observers during late July/early August.

Before moving on to the more complex and subtle lunar motions, we should mention the phenomenon known colloquially as the *harvest moon*. If you carefully follow the motion of the Moon against the background sky of fixed stars throughout a 24-hour period, you would discover that it appears to slide "backwards" (eastward) by about 13°. Were the Moon's orbit coincident with the celestial equator, it would therefore rise a little less than one hour later each successive night (the Earth turns at the rate of 15°/hour). This is what astronomers call *retardation*. But, as we have seen, the Moon does not orbit on the celestial equator, but at an angle inclined to the ecliptic, so that at different times of year its orbit makes different angles with the horizon. This in turn causes variations in the retardation, which average 50 min, 30 s.

As we have also seen, the Moon rises at different points along the horizon as the year goes by – on September 21, the time of the autumnal equinox, the entire southern section of the ecliptic lies above the horizon at sunset, with a very shallow angle between the two. Hence the Moon, in moving its usual 13° eastward each night at this time of year, has a smaller vertical distance to travel than during the spring equinox, when the ecliptic is sharply inclined to the horizon. As a result, the Moon around full, from a few nights before the autumnal equinox until a few nights after, appears to rise at the same convenient time every night. For as long as farmers in the northern hemisphere have struggled to pull in their harvests every September they have appreciated this phenomenon of the Moon's motion, calling it the harvest moon as it gives them a little extra light to work by, conveniently delivered just as the Sun sets.

Lunar Librations and Eclipses

If you already have an interest in astronomy, you have probably heard people use the term "libration" in connection with the Moon, but do you really know what it means? Answering that question is not as straightforward as you might guess, for the Moon has four different motions that may all properly be described as librations, all consequences of different kinds of movement. You might assume that, because the Moon always keeps the same hemisphere turned toward the Earth, as a consequence of its captured rotation, we always see the same 50% of the Moon's globe. But we actually get to observe, at one time or another, 59% of the lunar surface: a little bit "around" each of the east and west limbs, and a little bit "over" the north and south poles. How is this possible?

The first type of libration is called *libration in longitude*, and consists of a side-to-side wobbling of the Moon, like someone shaking their head to say "no." Although the Moon rotates on its axis at a constant speed, its orbital velocity around the Earth varies, as does that of any planet obeying Kepler's second law: the Moon moves faster when it's nearest the Earth (at perigee), but more slowly when farthest from us (at apogee). As the Moon moves towards apogee, it slows down in its orbit, so that its rotational speed exceeds its orbital motion. At this time we get to peek around the "normal" west limb of the Moon to see regions of longitude we ordinarily wouldn't be able to see. The opposite happens when the Moon moves towards perigee – its orbital speed overtakes its rotation, so we get to peek beyond the "normal" east limb of the Moon. Figure 1.5 illustrates libration in longitude.

Galileo deserves the credit for discovering the Moon's libration in longitude, which he first announced in 1632. He accomplished the feat by measuring the changing distances from the limb

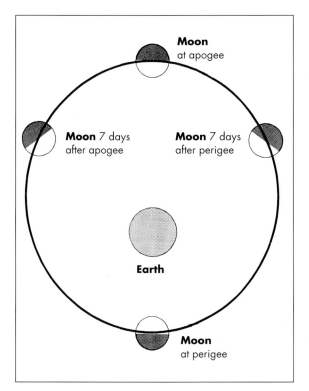

Figure 1.5. The Moon's libration in longitude. The Moon's orbit around the Earth is shown as an exaggerated ellipse, and the sizes of the Earth and Moon are exaggerated for clarity.

of Mare Crisium, one of the lunar "seas," and Grimaldi, a large crater, both of which lie near the lunar equator. This type of libration is of great interest to lunar observers, for it allows them to get a better look at features that are otherwise more greatly foreshortened by their closeness to the "normal" limb of the nearside Moon, as well as glimpses of other features that can be seen only at these times.

The second kind of libration is an up-and-down rocking motion of the Moon, like someone nodding their head to say "yes." Two factors combine to cause this *libration in latitude*: the 5° inclination of the Moon's orbital plane to the ecliptic, and the inclination of the Moon's rotational axis to its orbital plane (about 1°.5). As a result, the north pole of the Moon sometimes tips toward the Earth by as much as 6° 41′; at these times we can see beyond the north pole. Halfway around its orbit, the north pole tips away from the Earth by the same amount, bringing regions beyond the south lunar pole into view. Galileo also apparently discovered this libration, which he communicated in a letter written in 1637 to a colleague. As with libration in longitude, this rocking motion allows the enthusiastic lunar observer to study regions that would otherwise remain out of view. Figure 1.6 shows the region of the Moon near its south pole, with libration features visible at the extreme top (astronomical telescopes invert the image).

Figure 1.6. The lunar south pole, with libration features visible at the extreme top. *Photograph by Bob Levitt.*

The third kind of libration is a parallax effect, caused by the changing position from which an observer on Earth sees the Moon as we rotate on our own axis; it is termed *diurnal libration*. This libration permits us to see a little way round the west limb of the Moon as it rises, and a little way round the east limb as it sets. Diurnal libration amounts to slightly less than 1°, but is not very useful to the lunar observer, because it's not profitable to try to observe the Moon when it's rising or setting – the thick layers of atmosphere you are forced to look through will completely ruin the view by grossly distorting any features you are trying to see.

And lastly, because the Moon is not perfectly spherical, with a slight bulging pointing toward the Earth, there is yet another, albeit extremely small, effect known as *physical libration*. This manifests itself as a very slight variation in the rate of the Moon's rotation about its own axis, but since this motion amounts to only 0.5″ at the Moon's center, lunar observers will never notice it!

Before moving on to the next chapter, we should briefly discuss eclipses of the Sun and Moon, which are two of the more aesthetically pleasing spectacles that the world of astronomy has to offer. Eclipses of the Sun occur when the Moon perfectly interposes itself between the Sun and our planet, allowing us to observe structures of the Sun that are invisible at other times. Every year and a half or so, thousands of amateur astronomers from all over the world spend very substantial amounts of money to travel to often exotic locales in order to enjoy a total solar eclipse. This is a time-honored tradition that dates back to the nineteenth century, when it was very popular for major observatories to spare no expense in organizing elaborate eclipse "expeditions," often involving the transport of unimaginable amounts of complicated astronomical equipment to far-off destinations. The efforts of both historical and contemporary "eclipse chasers" to place themselves in the Moon's shadow is perhaps still more surprising when you consider that totality, the length of time that the Sun's disk is completely covered, lasts only a few minutes!

Annular eclipses are less interesting to astronomers because the apparent size of the Moon is not large enough to completely block out the glare of the Sun, so the corona and prominences cannot be seen. They are still a pretty sight, but not nearly as dramatic as a total solar eclipse. Figure 1.7 is a photograph of the 1988 total solar eclipse viewed from Borneo; Fig. 1.8 is a photograph of the 1994 annular solar eclipse observed from Illinois. Compare the two photographs – Fig. 1.7 shows the corona and a few red prominences, but neither of these features are visible in Fig. 1.8. Solar eclipses are covered in wonderful detail in another book in this series, *The Sun in Eclipse* by Michael Maunder and Patrick Moore.

A lunar eclipse happens when the Earth's shadow falls upon the Moon's nearside, often coloring the Moon an eerie red-orange or yellowish hue. Unlike solar eclipses, which occur at new moon, lunar eclipses can take place only at full moon and, like

Figure 1.7. The 1988 total solar eclipse as seen from Borneo, showing the corona and a few red prominences. *Photograph by Bernard Wahl.*

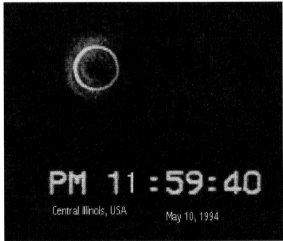

Figure 1.8. The 1994 annular solar eclipse as observed from near Urbana, Illinois. *Photograph by Bernard Wahl.*

solar eclipses, they require our satellite to be at or near a node (another reason for knowing your nodes!). We can actually experience three different kinds of lunar eclipse – total, partial, and penumbral – because the Earth's umbral shadow (the darkest, most concentrated part) extends beyond the Moon's orbit.

We are treated to a total lunar eclipse when the Moon is swallowed up by the umbra, during which time the Moon can become so dark as to almost disappear, especially if it is near the horizon or if there is smog in the air. A penumbral eclipse happens when the lighter part of the Earth's shadow, which is much wider than the umbra, falls upon the Moon. In contrast to total lunar eclipses, the effect of a penumbral eclipse is often almost unnoticeable, a slight darkening of the lunar disk the only telltale sign of the event. A lunar eclipse is termed partial when only part of the umbra obscures the Moon. We shall look at lunar eclipses in much greater detail in Chapter 11, "Lunar Eclipses and Occultations."

Chapter 2
The Basic Kinds of Lunar Formations

The heart of this book is found in Chapters 4 through 7, which explore the four quadrants of the Moon in great detail, describing the most interesting formations for you to subject to the scrutiny of your telescope, camera, video camera or CCD. These chapters are illustrated by drawings of selected features, all made by amateur astronomers. If you are not familiar with the nomenclature of the Moon's topography, this chapter will quickly introduce you to the basic kinds of lunar formations, with some examples. These include craters of various types (terraced, concentric, ghost and ray craters, crater chains), the lunar maria and bays, wrinkle ridges, rilles or clefts, mountains (ranges and isolated peaks) and lunar domes.

The Lunar Craters

The word "crater" derives from the Greek word κρατερ, meaning "cup" or "bowl." There also happens to be a constellation in the starry heavens called Crater – and not surprisingly, it represents a cup. The word is well chosen, for I can't think of a better way to describe these walled enclosures that dominate the lunar surface like no other topographic feature. When we think of the Moon, we think of craters. They are essentially impact sites – huge, shallow holes dug from the lunar surface by asteroids, meteoroids or perhaps comets that wreaked their havoc mostly billions of years ago during a chaotic period of lunar history known as the heavy bombardments. Because craters are surrounded by walls of debris, they are also known as *walled plains* or *ring plains*, terms now regarded as a little old-fashioned by the professionals, but still used by amateur lunar observers. While you are no

doubt familiar with the appearance of craters, have you ever given any thought to just exactly what these structures are, how they form, or whether they are all the same or different?

The Moon's craters are certainly ubiquitous – if we limit ourselves to craters of 1 km (0.6 mile) or more in diameter, we discover that there are three hundred thousand of them, and that's just on the Earth-facing hemisphere of our natural satellite! Figure 2.1 is a typical lunar scene – the heavily cratered landscape around the large craters Maginus and Tycho. The craters range in size from small pits a few meters across to giant impact basins hundreds of kilometers wide. One of the first things you should note about the size distribution of lunar craters is that their numbers increase greatly with diminishing size. This is consistent with what we would expect if craters were formed by impacts, and I shall have more to say about this later.

Craters are truly found everywhere we look, randomly distributed over the lunar surface, with one notable exception. Many of the mare areas were flooded with lavas that drowned most or all of the older craters within them, so that the deepest parts of these solidified lava flows, usually in the central region of the mare, are conspicuously devoid of craters. Most craters found within a mare will either be at the outer margins, where the ground was high enough to keep the crater above the surging lava, or, if away from the edges, they will be very young, fresh *craterlets* (very small craters). I shall talk more about the maria, which are just big, flooded craters , later in this chapter. But nothing has been able to stop the cratering process on the Moon – we even find craters atop lunar mountains and crater rims, and it is not unusual to find that a crater includes many smaller craters or craterlets within it.

Figure 2.1.   The heavily cratered landscape around the craters Maginus and Tycho. *CCD image by Maurizio Di Sciullo.*

The International Astronomical Union has recognized official names for hundreds of the largest craters. They are given the names of astronomers, other scientists, explorers and other famous persons, including political and religious figures from the past. In coming up with this nomenclature, special consideration has been given to astronomers who made contributions to our knowledge of the Moon, which is only fitting.

Craters have three or four main components. Their defining characteristic are their *walls* or *ramparts*, which in turn are made up of three separate constituents: an outer slope, or *glacis*, an inner slope, which often displays structures called *terraces* that are the result of great landslips, and the *rim* or *crest*, which often has peaks. In fact, the mountains of the Moon are for the most part just the rims of impact craters.

If you could look down on almost any lunar crater from above, you would notice that the outline of the walls would be essentially circular. But from Earth, many craters look very oval or elongated; these tend

to be located near the limbs of the Moon, and look the way they do because of the foreshortening effect, briefly described in Chapter 1. If we correct for the foreshortening effect, we realize that these craters are round also. This is not to say that crater walls all make a perfect circle – it's not unusual for the ramparts to be broken or offset, or to have valleys or canyons cutting through them. The lunar terrain varies from one impact site to another, and faults and other weaknesses in the crust have shown themselves by causing crater walls to be offset in a characteristic polygonal shape. Some crater walls are mostly circular, but with one or two "straight" segments where faulting has occurred.

The walls of other lunar craters take on a braided appearance that commentators describe as "ropy," a consequence of the blanket of *ejecta*, material ejected by the crater-forming impacts, laid down around them. Many crater walls have been disturbed by smaller craters that formed later, but these minor pockmarks are nothing compared with the destruction that lava flows have wreaked, knocking down whole sections of the wall and flooding the floor. Despite all of these factors, most crater outlines are nevertheless quite round.

The roundness of lunar craters might surprise you, as it did many of the astronomers who studied these features in the first part of the twentieth century. They reasoned that if meteoroid impacts were responsible for most, if not all, of the Moon's craters, then we should see lots of craters that are truly oblong (even when foreshortening is corrected for) because meteoroids would be expected to strike the lunar surface at all kinds of angles, including some very shallow angles that would tend to produce ovals instead of circles. In fact, the absence of truly oval lunar craters was sometimes used to support the notion that lunar craters were actually volcanic in origin, since terrestrial volcanic craters all tend to be fairly circular.

But this problem was solved by Thomas Gold in an important paper published in 1956 which built on the work of Ralph Baldwin and other scientists. Gold showed that impacting meteoroids are moving with such high velocities (typically many kilometers per second) when they strike the Moon that the explosion responsible for carving out the crater acts essentially like a "point source," which digs out a circular hole no matter what its strike angle is. Actually this is an oversimplification, for impactors striking the Moon at extremely shallow angles *could* produce an elongated gouge, but for the most part Gold's calculations explain why almost all lunar craters are fairly round. He showed that even small impactors are capable of creating large craters, 25 to 50 times their diameters, because their huge kinetic energies are released upon impact and converted into mechanical, thermal, and acoustic energy. The impacting meteoroid penetrates the lunar surface where it is swallowed up in less than a second, sending powerful shock waves through the lunar crust. Some of these

seismic waves travel back up through the meteoroid, which is partly vaporized during the impact. Think of the seismic shock waves as the Moon's attempt to dissipate the forces of impact. These attempts are inadequate, however, for the meteoroid strikes the Moon with a velocity 3 to 4 times greater than the speed at which these seismic waves can travel outwards, typically 5 km s^{-1} (3 mile s^{-1}), so that the energy must go somewhere else.

Where does it go? Well, for one thing, both the impactor and the target lunar rock are violently compressed, and the shock and decompression waves (whatever is compressed must *de*compress!) dig out rock in the characteristic bowl shape, throwing up the excavated rock into a ring around the bowl, forming the crater walls and ejecta blanket (the walls, which are what we recognize as lunar mountains, are also created in part by the shock waves directly). Some lumps of the ejected material are bigger than others, and the larger pieces of rock are capable of making their own, smaller *secondary craters*. Sometimes huge blocks of rock blown clear of the impact site form deep grooves in the lunar surface that radiate away from the impact basin, while lighter ejecta is blasted into space to fall gently back to the lunar surface, forming a *ray system*. Finally, the floor of the crater tends to rebound from the initial shock, a process which often results in instant mountain-building at the center of the crater, forming a so-called *central peak*. It is possible that some of the fractures we see on the floors of lunar craters are also artifacts of the initial impact, though if these fractures were deep enough, they would have released lavas from below that would have subsequently erased evidence of their existence. Figure 2.2 shows Gold's model for crater formation, which is still accepted today.

Aside from their general outlines, not all crater walls look alike, and differences in their appearance

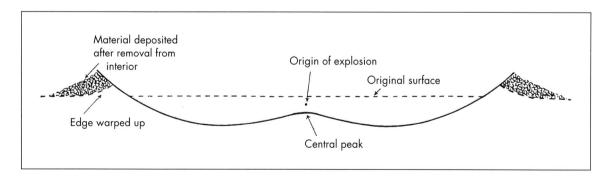

Figure 2.2. Thomas Gold's model for crater formation. *Adapted from Gold, T, The Lunar Surface. Monthly Notices of the Royal Astronomical Society, 115:585–604, 1955.*

give clues about the geologic history of the crater, especially its age. Younger craters tend to have very sharp crests that have not been broken down by "erosion." Before going any further, we should carefully define this potentially misleading term, which appears in many lunar guidebooks. When we speak of erosion on Earth, we are mainly referring to the long-term effects of water and weather. But, as we know, the Moon has neither running water (the water ice that is likely trapped inside some lunar rocks, especially at the south pole, never melts and flows across the surface, and has no erosive effect) nor an appreciable atmosphere to support the weather we experience on our planet. In fact, this is one reason why we find so many more craters on the Moon than on Earth – billions of years of violent geologic and climatic processes, from plate tectonics, volcanoes and miscellaneous seismic events to the forces of oceans, rivers, rains and wind have all ganged up to obliterate or make nearly invisible the vast majority of Earth craters, so that today only a few well-preserved examples of relatively recent impact structures remain.

It should be mentioned that the Earth's thick atmosphere has also protected us from many impacts, especially smaller ones, by causing incoming meteoroids to burn themselves up by a process known as ablation. The Moon has no such protective atmosphere, so anything heading for the Moon will strike its surface. Figure 2.3 is a photograph of the interior of Meteor Crater in Arizona, the most famous example of an impact site on our planet. The Moon has experienced few of the abovementioned climatic and geologic processes throughout its 4.5 billion year history,

so its face has changed far less than Earth's over that period of time. The lunar landscape is ancient indeed, so that when you look at its highly cratered face you are in a sense looking backwards in time.

But if the Moon is not subject to the weathering processes we see here on Earth, why are some craters more eroded than others? The answer is that on the Moon erosion takes place somewhat differently. Oddly enough, one of the main erosive forces on the Moon is probably the same process that created the craters in the first place – impacts. Most of the craters we see have suffered subsequent, usually smaller, impacts. It is easy to see many examples of later cratering that alter the appearances of craters that were there before. Each impact pulverizes the lunar rock it strikes, and if the impact is energetic enough it can also melt and reform this lunar rock; this phenomenon is discussed further in Chapter 3. Even the tiny, so-called micrometeorites that are constantly striking the Moon (and the Earth!) in great numbers have a long-term weathering effect on the lunar soil, gradually changing its texture and color – a process known as *gardening*.

We now know that lava flows occurred on the Moon billions of years ago, though they resulted from processes that are different from the eruptions of discrete volcanoes typically found on Earth. These were gentle "sheet" flows that covered huge areas and persisted for perhaps hundreds of millions of years. This process has also had an "erosive" effect on lunar craters, and in general it is correct to conclude that craters with lava-flooded floors or broken-down walls (often the two go hand in hand) are older than craters that do not bear the scars of these affronts.

Figure 2.3. The interior of Meteor Crater in Arizona. *Photograph by the author.*

Besides sharply defined walls, younger lunar craters also tend to show rich terracing on their inner slopes, as well as central mountain peaks. It is probable that many older craters would show these features too, but they are buried under the lavas that fill their interiors. On the other hand, we can find examples of many older craters whose floors are *not* flooded but which also look broken down and eroded, with jumbled, rounded walls that show little or no terracing, and central peaks that are either broken down or missing altogether. These craters have probably been battered by later impacts and perhaps further collapsed by seismic events – even a large impact on the opposite side of the Moon can cause unimaginably violent moonquakes capable of shaking craters up.

All craters also have a *floor*, which may be rough or smooth, depending on what geologic or impact processes have taken place inside the crater since its formation. The impact itself will fracture the floor's bedrock, and usually leaves other telltale signs, like rubble piles or pools of resolidified melted rock. A smooth floor is usually an indication that lavas have resurfaced the crater's interior. Sometimes debris ejected by a neighboring crater will scour the floor of an already-existing crater. You can see a good example of this process on the floor of the crater Ptolemaeus. It is interesting that crater floors are almost always lower than the level of the terrain outside the crater's walls. What is more interesting, the degree to which a crater's floor is depressed below the surface is related to the crater's diameter – the larger the crater, the more depressed its floor is likely to be. This is another clue that lunar craters are produced by impacts: the more energetic an impact, the deeper and broader the resulting crater will be. Indeed, when you examine most lunar craters using a telescope, they look like very deep bowls, and it is not unusual for a crater's walls to rise 3000 m (10 000 ft) or more above the floor, even though these same walls may be only half that altitude above the exterior ground.

One thing you should keep in mind, and which may surprise you to know, is that even though it is common for a crater's rim to crest thousands of meters above its floor, this height tends to be very small compared to the crater's diameter. In other words, craters tend to be surprisingly shallow. Figure 2.4 shows the crater Clavius and its actual depth profile. When you look at this photograph of the crater, which closely simulates what you would see in the eyepiece of your telescope, it looks extremely deep, doesn't it? But when you look at the plotted profile it is obvious that Clavius, despite its telescopic appearance, is actually quite shallow.

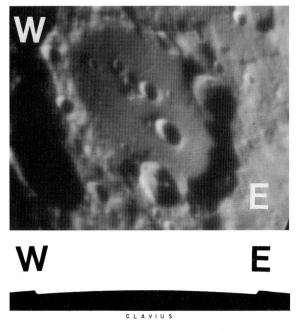

Figure 2.4. The crater Clavius and its depth profile, showing corresponding points on the crater's western and eastern rims. *Photograph by Bill Dembowski.*

While the inner walls of a typical lunar crater may slope by as much as 30°, the outer glacis often slopes much more gently – typically by 15° or less. The deepest lunar craters have depths equal to one-fifth of their diameters; these tend to be smaller craters. The larger craters are shallower – only one-thirtieth as deep as they are wide.

We detect topographic relief on the Moon visually through the shadows cast by the Sun's rays as they strike features such as a crater's walls at an oblique angle. The smaller the Sun angle, the better we see the relief. When the Sun is directly over a feature on the Moon, the feature looks completely flat – it loses its relief. Conversely, when the Sun shines on a feature almost sideways, the feature, if it is a crater, will look very deep. Our eyes tend to overestimate the true depth of lunar craters, thanks to the dramatic shadows cast by the Sun, which with no atmosphere to diffuse them, are extremely sharp. As we shall see in Chapter 10, these shadows are extremely useful as they allowing us to calculate the heights of lunar mountains and crater rims.

Most craters also have a group of hills or mountains, or at least an isolated peak, at the center of their floors. These features, like the walls, are relics of the impact process, and were explained by Baldwin, Gold and others in the mid-twentieth century. Earlier researchers tended to assume that these central peaks were volcanic cinder cones, but decades of geologic research, including high-resolution images obtained by spacecraft and analysis of lunar samples returned to Earth by the Apollo astronauts, have shown that these features are not volcanoes, just modestly sized mountains formed by the impact process itself.

It was not entirely unreasonable for earlier lunar scientists to believe that central peaks were volcanoes, since many of them, like the peak found at the center of the crater Regiomontanus, are indeed capped by craterlets that mimic the appearance of volcanic calderas, and moreover, as we have seen, many crater floors have been flooded by lava. The larger craters tend to have a group of central peaks, while smaller craters usually have only a single central peak. These peaks are not terribly high – usually they're about 900–1200 m (3000–4000 ft). Sometimes you will see a ring of mountains on a crater's floor – this is an indicator that another impact occurred inside the main crater, the mountain ring evidence of the smaller crater's rim.

Besides being of different ages, craters come in different varieties. A *terraced crater* is, as we have already mentioned, just a very well-preserved crater that shows concentric rocky ledges along its inner walls. A good example of this type of crater is Copernicus, shown in Fig. 2.5. As we have seen, the Moon is so heavily cratered that virtually all sizable craters have other, smaller craters within their walls. Every so often, a later impactor will aim itself right

at the heart of an existing crater, gouging out a second crater inside the first, but a little smaller, whose rim is perfectly concentric with the outer walls of the older crater. An example of such a *concentric crater* is Hesiodus.

If lavas have so completely flooded over an already existing crater as to render it almost, but not completely, invisible, we call the remnants of the crater rims that peek above the lava plains a *ghost crater* or *ruined crater*. If only part of the crater wall has been broken down by lava flows, we refer to the structure as a *partly ruined crater*. A good example of a partly ruined crater is Fracastorius, shown in Fig. 2.6: its whole north wall has been overwhelmed by lava flows from Mare Nectaris. If a partly ruined crater is large enough and situated at the margins of one of the lunar maria ("seas") it may colloquially be referred to as a *bay*.

A *bowl crater* is just a small crater, 20 km (12 miles) or less in diameter, that is nothing more than a concave hole in the ground. The impacts that form bowl craters are not energetic enough to create high rims or central peaks. Bowl craters are thus smooth and featureless.

Some of the freshest impact sites on the Moon are *ray craters*. If you look hard enough, you will notice some evidence of debris ejected by the impact around most lunar craters. This debris may take the form of a milky while *ejecta blanket* dusting the outer slopes of the crater walls, or perhaps less organized piles of rubble inside or outside the walls. The ejecta blanket often takes on a hummocky appearance and can spread out for up to one crater diameter. A few craters radiate several *rays* of ejecta material. Lunar rays show no shadows, so we

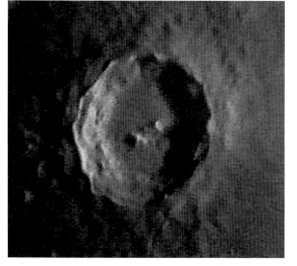

Figure 2.5. The crater Copernicus, an example of a concentric crater. *Video image by Joe LaVigne.*

Figure 2.6. Fracastorius, a lava-flooded ruined crater. *Video image by Joe LaVigne.*

know they have no appreciable height or thickness. They are just splashes of finely pulverized lunar rock (sometimes called *rock flour*) sprayed across the surface of the Moon for up to hundreds of kilometers. Often rays spread out in all directions from their parent crater, forming patterns that have been likened to a peeled orange or a fractured Christmas ornament. In general, rays are best seen around full moon; rays associated with a particular formation may be seen when the Sun is overhead at that formation. Use binoculars rather than a telescope to observe them, for these ray systems extend across large amounts of ground.

Lunar rays formed when pulverized particles of rock flew away from the impact that created the associated crater. Much of this material was accelerated to speeds of up to 40 km s^{-1} (25 mile s^{-1}), much greater than the Moon's escape velocity; some slower-moving debris falls back to the lunar surface to create the typical ray pattern. Craters like Tycho and Kepler have very bright, extensive ray systems. Probably many more craters showed rays at one time, but most of them have faded away over time due to the discoloring effect of ions that make up the solar wind and as a result of the gardening action of micrometeorites, which tends to mix up the whitish ray material with the darker underlying lunar soil. Ray systems are further evidence for the impact origin of lunar craters – during World War II, scientists discovered that bombs produced craters with ray systems just like those found on the Moon. There are many unsolved mysteries surrounding the lunar rays which planetary geologists are trying hard to solve. For example, why is it that some fairly inconspicuous craters have huge ray systems, while the rays of much more impressive craters are much less spectacular? Why do some ray systems seem to have multiple foci? Why do their shapes vary so much from one ray system to the next?

The lunar surface is also covered by many *crater chains*, which probably come from two different kinds of impact event. Some very short crater chains, which are probably better termed "crater groupings," are chance alignments of a small number of impact sites. The walls of these grouped craters often overlap or obliterate one another. Secondary craters often group together like this. Other crater chains are the likely result of a single impact event, created by a skipping or rolling impactor that gouged its way across the lunar soil in a straight, or nearly straight, line. Some of these gouges form a herringbone pattern, with the "V" pointing back towards the culprit crater.

The last variety of craters we need to mention are the truly *volcanic craters*, which are vents at the

heads of lava tubes. The best-known example is the oblong crater known as the "Cobra's Head" lying at the likely source of the collapsed lava tube known as Schröter's Valley, near the craters Aristarchus and Herodotus. Yes, the Moon has experienced volcanic activity in its distant past, but this activity accounts for only a very insignificant number of the hundreds of thousands of lunar craters.

The Lunar Maria and Bays

The lunar maria, or "seas" (so-called because ancient Moon-gazers speculated that they were analogous to Earth's oceans) are gigantic circular impact sites that have filled with volcanic magma that has long since cooled into a wide variety of basaltic rocks. In binoculars, many of them look like big craters – Mare Crisium near the Moon's east limb is a fine example. They often have rings of mountains around them. The principal maria are Mare Frigoris (the Frigid Sea), Mare Imbrium (the Sea of Rains), and Mare Serenitatis (the Sea of Serenity) in the northern regions of the Moon; Oceanus Procellarum (the Ocean of Storms), Mare Vaporum (the Sea of Vapors), Mare Tranquillitatis (the Sea of Tranquility), Mare Crisium (the Sea of Crises), and Mare Fecunditatis (the Sea of Plenty) in the Moon's equatorial regions; and Mare Humorum (the Sea of Moisture), Mare Nubium (the Sea of Clouds), and Mare Nectaris (the Sea of Nectar) in the southern regions.

Planetary geologists have discovered that giant, asteroid-sized impactors capable of digging out impact basins of at least 300 km (200 miles) in diameter would have carried sufficient energies to create what are known as *multi-ringed impact basins*. As their name implies, these are huge bull's-eye-shaped concentric rings on the lunar surface, the best example of which is the Mare Orientale basin, shown in Fig. 2.7 just around the west side of the Moon's limb. (Before the International Astronomical Union redefined directions on the Moon to match those used by astronomers for the sky as a whole, east and west used to be the other way round, so that Mare Orientale, the Eastern Sea, is now found on the western instead of the eastern limb!) After the discovery that Mare Orientale actually had another impact ring around its more obvious "shores", lunar scientists re-examined the nearside maria, discovering that they too were multi-ringed impact basins. A good example may be seen surrounding Mare

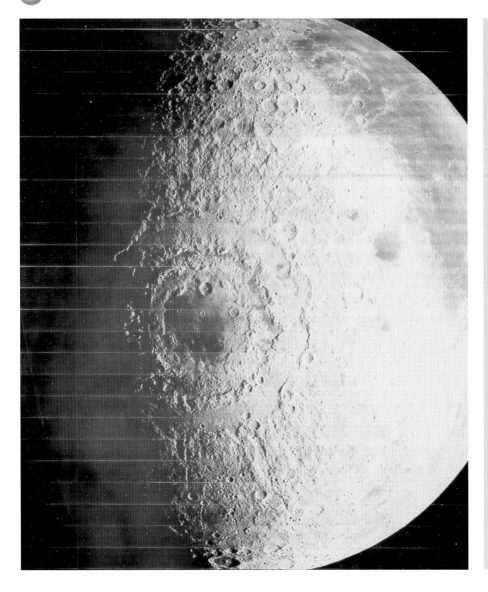

Figure 2.7.
Mare Orientale
imaged by Lunar
Orbiter. *Courtesy
Lunar and Planetary
Institute (LPI),
Houston, Texas.*

Nectaris, where the Altai Mountains form a partial secondary ring around the more obvious dark, lava-filled halo of Nectaris's floor.

The mare areas are dark because they flooded with basalt lavas that were once highly fluid, before they cooled, flowing downhill and filling the lowest areas first. Look at any mare region through a pair of binoculars to prove this to yourself – the central portions are very smooth and dark, generally devoid of any major topographic features. This is where the impact basin, and therefore the lava filling, was deepest. But around the margins you will begin to see a few craters and other topographic features that were on higher ground and which at least partially survived the filling process. The lava flows on the Moon took perhaps hundreds of millions of years – there is evidence the lavas were laid down in very

thin sheets by a very gradual process, very unlike the Earth's brief and violent volcanic eruptions with which we are more familiar.

Along the "shores" of many maria you will find the lunar bays. Figure 2.8 shows Sinus Iridum, the colorfully named Bay of Rainbows, located at the northern edge of Mare Imbrium. Another well-known example of a lunar bay is Fracastorius, on the southern shore of Mare Nectaris. The lunar bays are nothing more than secondary impact sites indenting the surrounding highlands, whose walls have been partially obliterated by the basalt flows from the maria that harbor them.

Also around the margins of the lunar seas we encounter the wonderfully enigmatic features known as *wrinkle ridges*, the largest of the Moon's tectonic features. Wrinkle ridges really do look like

Figure 2.8. Sinus Iridum, the Bay of Rainbows. *CCD image by Ted Wolfe.*

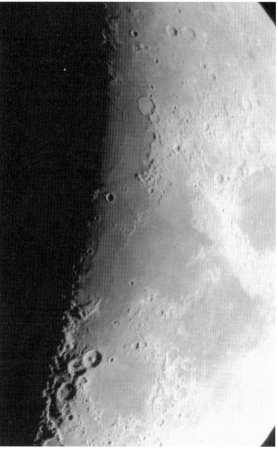

Figure 2.9. Wrinkle ridges in Mare Serenitatis, including the famous Serpentine Ridge (just above center, running northward (up) from Pitatus, the small, "sharp" crater just left of center), and Mare Tranquillitatis. *Photograph by Bob Levitt.*

wrinkles in the Moon's crust, giving it the appearance of a pudding skin. You will see prominent examples of these features in almost every mare basin, but especially crossing the eastern third of Mare Serenitatis and along the western half of Mare Tranquillitatis. They are best observed at grazing Sun angles, which tells us that they are not terribly elevated above the mare plains; most are only a hundred or so meters high. Figure 2.9 shows wrinkle ridges in Mare Serenitatis (the famous "Serpentine Ridge") and Mare Tranquillitatis.

The wrinkle ridges tend to be concentric with the circular borders of the maria themselves; occasionally they replace the missing walls of lunar bays, as with Sinus Iridum or Letronne. In other cases they appear to outline older, probably submerged impact basins or large craters. Many wrinkle ridges tend to run north and south for great distances, and in binoculars may look fairly straight. But if you examine them more closely, you will discover that they are often sinuous and braided, rather like twisted pretzels. Because they occur in so many different circumstances and take on different appearances depending on where they occur, planetary scientists have proposed a number of different theories to explain what wrinkle ridges are and how they formed.

One of the more popular is the subsidence theory, which says that wrinkle ridges formed after mare lava solidified at levels higher than their current heights, eventually sinking down to make the ridges by compression as the mare surface area decreased – simply folds in the cooling lavas. There is evidence of this subsidence at the margins of the maria, where we find many wrinkle ridges. But how does this explain the wrinkle ridges in Mare Imbrium, which betray the inner ring of a multi-ringed impact basin? Here the wrinkle ridges look like nothing more than the submerged walls of a very large impact site. Another theory suggests that great fissures in the lunar crust allowed magma to well upwards, solidifying as it cooled along the roughly linear fissure lines. Other wrinkle ridges appear to be related to simple highland faults. Perhaps there really are different types of wrinkle ridge that require separate theories to account for them all, but for the time being they remain among the most confounding of lunar features.

The Lunar Rilles

Many of the Moon's impact sites, including both the expansive maria and the larger craters, are crossed by a wide variety of long, narrow valleys, gorges, furrows, and gullies, collectively referred to as *rilles* or (the old term) *clefts*. Most of these features are less than 2 km (about 1 mile) wide and are best observed at Sun angles of around 12°, which occurs one day after local lunar sunrise or one day before local sunset. Rilles tend to look like delicate cracks in the lunar surface, and in at least some instances that's probably just what they are. But it would be misleading to group all of these features together, for they include structures that almost certainly result from widely disparate processes. For example, some rilles are nothing more than the crater chains we have already mentioned, a relic of the impact process, while others are probably lava tubes produced by lunar volcanoes. So rilles tend to include features that, though they look similar to the casual observer, are not really related to one another. It is therefore helpful to adopt some sort of classification scheme so that we know what types of rilles we're talking about.

Sinuous rilles are so named because of their winding, serpentine shapes. Schröter's Valley is a sinuous rille – it is a collapsed lava tube that poured out from the volcanic vent called the Cobra's Head next to the craters Aristarchus and Herodotus, forming a rille that meanders, like a river, downhill through the highlands to the north of those craters, twisting and turning as it goes, finally fading away after it reaches level ground. These structures, which tend to be found on the surface of the maria, formed when a crust hardens over a molten river of lava, the lava underneath eventually draining away to leave a hollow tube. Sheer gravity or the blows of meteoritic impacts might cause the upper crust to collapse into the bottom of the tube, creating a channel not unlike river beds on Earth. Geologists are very familiar with lava tubes on our planet, and there is much about the terrestrial variety that is morphologically similar to structures like Schröter's Valley.

Hadley Rille is another fine example of a sinuous rille. The astronauts of Apollo 15 explored the landscape around Hadley Rille using the famous Lunar Rover, and got some very dramatic pictures of it from ground level. Located at the bottom of the Apennine Mountains, Hadley Rille also shows the bends of a typical terrestrial river. At one time, some

astronomers actually thought that sinuous rilles might have been carved by running water; today, our improved knowledge of the Moon's geologic history, discussed in the next chapter, tells us that no such thing ever happened. For example, the rocks from Hadley Rille collected by the Apollo 15 crew should have showed some evidence of water if it were a dried-up river bed, but they did not.

Straight rilles, which include *graben-type rilles*, are completely different from sinuous rilles. They are, as their name implies, much straighter than sinuous rilles and, unlike the former, they tend not to be confined to the mare lava plains. Straight rilles are probably related to faulting mechanisms in the lunar crust. But we must keep in mind that the Moon shows little evidence of the plate tectonics that are so important in shaping the Earth's topography – there are none of the rift valleys or displacements across ancient craters that we would expect to see if plate tectonics were at work there. So when we talk of "faults" on the Moon we are talking about more modest processes: either a simple crack in the surface, perhaps caused by the stresses of impact, as in the case of the Sirsalis Rille; or perhaps a process by which the surface crust contracts a little, allowing a length of ground caught in the middle of the receding sides of the fault line to slump down a bit, creating a wide, shallow graben-rille. The best example of such a rille is the Alpine Valley.

The third kind of rille, known as the *arcuate rilles*, are similar in many respects to the straight rilles, but they tend to be found near the shores of the lunar maria, where dense basalts sank from their own weight in a process known as isostasy, whereby heavier material sinks until it reaches a depth at which the upward pressure is sufficient to support them, while lighter materials rise until they reach a similar equilibrium state. The arcuate rilles, so called because they tend to follow the arcs of the mare shorelines, probably caused the solidified lava at the mare margins to crack under its own weight – we can see excellent examples of such rilles around the edges of the Mare Humorum and Mare Serenitatis basins.

As with the wrinkle ridges, there are many mysteries associated with lunar rilles. For example, one of the most famous rilles, the Hyginus Rille (see Fig. 2.10) is difficult to explain under any current theory of rille formation. As you can plainly see, it is peppered with tiny craters along its length, but it truly appears to be a rille, not just a crater chain. Some lunar geologists believe that the many craters along the bottom of the Hyginus Rille suggest a volcanic origin for this feature.

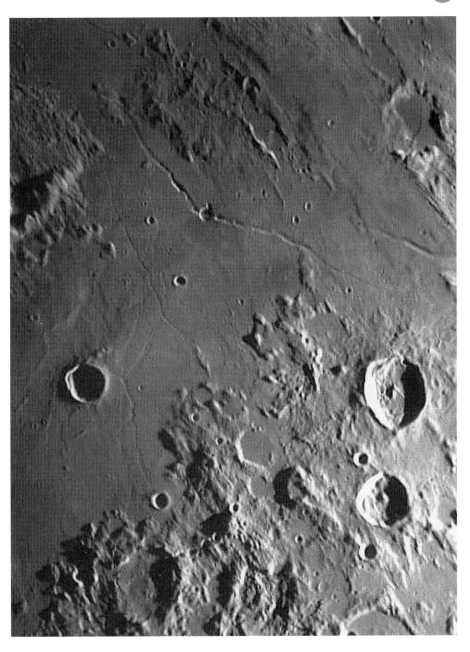

Figure 2.10.
The Hyginus Rille.
CCD image by
Maurizio Di Sciullo.

The Lunar Mountains

There are sixteen major *mountain ranges* and huge numbers of *isolated peaks* highlighting the lunar surface. Lunar mountain ranges are named for Earthly counterparts, despite only a bare morphological resemblance because, as we have seen, they are formed by impacts, not tectonics. In almost every case we examine, lunar mountains constitute the encircling walls of impact sites. Many lunar mountain ranges form the boundaries of large

impact basins, for example the Haemus Mountains that fringe the southwest side of Mare Serenitatis, or the Carpathian Mountains that make up the southern rim of Mare Imbrium.

Not all high lunar mountains are found in ranges – some are lonely sentinels on the mare surface, stranded far from any neighboring highlands. Perhaps the best known example of an isolated peak is Pico, in Mare Imbrium south of the crater Plato, which rises to a height of 2500 m (8000 ft). Other examples of isolated peaks are La Hire (1500 m, 5000 ft) and Mt Piton (2000 m, 7000 ft) in Mare

Imbrium. If the lunar highlands are analogous to the geographic features we call "capes", they are also referred to by this name, or alternately as *promontories*. Two good examples of promontories are Cape Laplace and Cape Heraclides, which guard the entrance to Sinus Iridum. They are just the bluffs on either end of the mountain ring surrounding the Iridum impact site where this ring is broken off by the lavas that have intruded from Iridum itself.

You may well ask, how high are the mountains of the Moon? This is a very interesting question, only briefly answered in Chapter 1, where we discovered that 3000 m (10 000 ft) high crater rims are not unheard of. This, of course, is not a very complete answer to the question. Most crater rims are a few thousand meters high, but peaks in the major ranges are higher still – Mount Hadley, near the northern edge of the Apennines, is 4500 m (15 000 ft) in altitude, while to the east of the crater Calippus is a peak rising to 5800 m (19 000 ft), and though there are uncertainties because of their nearness to the Moon's south pole, some selenographers have discovered peaks in the Leibnitz range as high as 8800 m (29 000 ft)!

I have already stressed that, because of the Moon's small size relative to the Earth, these heights relative to the overall lunar topography are still more impressive than the absolute figures themselves suggest. But the fact of the matter is, we have reliable height measurements for only a very small fraction of the Moon's mountains and crater rims. It is possible for the amateur lunar enthusiast to make extremely valuable contributions to our knowledge in this area by using relatively simple equipment and specialized software to calculate the altitudes of lunar features for themselves.

Beginning in the northern regions of the Moon, as we did with our listing of the major mare basins, we find Mare Imbrium surrounded by several mountain ranges that make outstanding targets for the amateur lunar observer. The Apennines, already mentioned in this chapter, form much of the southeastern boundary of the Imbrium basin. There can be no better place to acquaint yourself with the mountains of the Moon, for the Apennines extend for almost 650 km (400 miles) from the crater Eratosthenes to the gap between Mare Imbrium and Mare Serenitatis.

Figure 2.11 is a fine CCD image of the Apennines, which contain no fewer than three thousand separate mountain peaks! Some of the noteworthy peaks in this range include Mt Hadley (which we've already described), Mt Wolff (3700 m, 12 000 ft), Mt Bradley (4250 m, 14 000 ft) and Mt Huygens (5500 m, 18 000 ft). Like many ranges that form the rims of mare impact basins, and for that matter like

the walls of the larger lunar craters, the inward-facing side of the Apennines is a steep curve of cliffs overlooking the mare floor, while the 240 km (150-mile) wide outward slopes are far gentler. The 290 km (180-mile) long Carpathian Mountains complete the southern wall of Mare Imbrium, stretching from the crater Tobias Mayer to Eratosthenes where they merge with the Apennines. The Carpathians are less contiguous than the Apennines, showing more signs of severe impact shock and subsequent erosion. They are also less lofty, their highest peak no more than 2000 m (7000 ft). The Caucasus Mountains form the eastern wall of the Imbrium plain. They are very rugged indeed – we have already mentioned the 5800 m (19 000 ft) high peak at the northern extreme of this range.

The lunar Alps border Imbrium to the northeast. The Alps are very broad, with hundreds of peaks having altitudes between 1500 and 2500 m (5000 and 8000 ft) spreading towards Mare Frigoris to the north. Mont Blanc, overlooking the shores of Mare Imbrium, rises to a height of 3500 m (12 000 ft). The Alps are bisected by the famous Alpine Valley, a huge scar from the Imbrium impact that radiates away from the center of the basin. Figure 2.12 is a fine image of the Alpine Valley taken with a CCD camera.

The northern margins of Imbrium harbor three small but interesting ranges – the Spitzbergen Mountains, the Teneriffe Mountains, and the appropriately named Straight Range. On the far western edge of the Imbrium plain you will find the Harbinger Mountains, with peaks of 2000 m (7000 ft) or so, near the craters Aristarchus and Herodotus. Before leaving Imbrium, we should mention the Jura Mountains, which form the rim of the Sinus Iridum. They reach a maximum altitude of 4500 m (15 000 ft) and are well preserved, with many lofty peaks.

Moving on to Mare Serenitatis, the Haemus Mountains are the primary range here, lying along the southwestern shores. This range offers many peaks with modest altitudes between 1200 and 2500 m (4000 and 8000 ft). The whole range appears very bright, but with few valleys separating its mountains. To the east of Mare Serenitatis lie the Taurus Mountains, a very broad highland region with mostly low mountains, but also a few peaks topping 3000 m, (10 000 ft).

The Pyrenees, located in the Moon's southern hemisphere along the east margin of Mare Nectaris, separate it from Mare Fecunditatis. They are over 300 km (190 miles) long, with numerous valleys separating the major peaks, at least one of which attains an altitude of 3500 m (12 000 ft). Opposite the Pyrenees is the important Altai Scarp, a mountain range that curves for 440 km (275 miles)

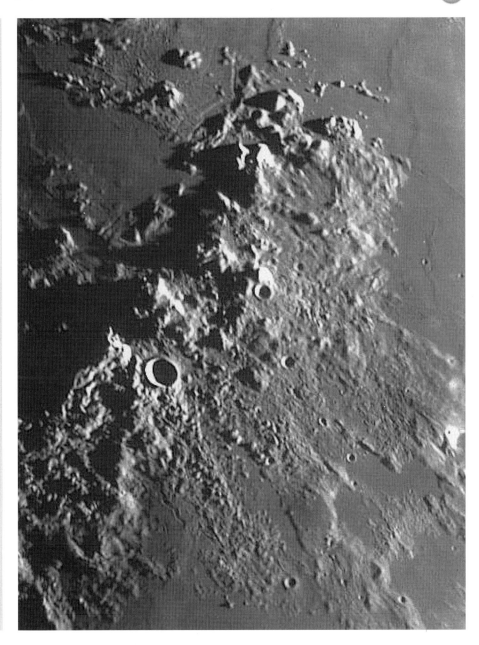

Figure 2.11.
The Apennine
mountain range.
CCD image by
Maurizio Di Sciullo.

concentric with, but far outside, the southwest shores of Mare Nectaris. The highest peak here is around 4000 m (13 000 ft). Starting from the Altai Scarp, it is possible, using binoculars or very low power on a telescope, to trace the remains of a second, outer impact ring encircling the Nectaris basin. Like Mare Orientale, Nectaris is clearly a multi-ringed impact basin.

There are many other mountain ranges worthy of study on the lunar surface – the Riphaen and Ural ranges in the extreme northwest hinterlands of Mare Nubium, a rather irregularly shaped complex of multiple impacts, and quite a few ranges that skirt the lunar limbs – besides the Leibnitz

Mountains atop the south pole there are the Rook, Cordillera and d'Alembert ranges on the west limb, peaks in the Rooks and d'Alemberts soaring to 6000 m (20 000 ft) or more. In the same category are the Dörfel Mountains, on the southwest limb, with peaks of similar heights.

The walls of large craters such as Copernicus and Arzachel may also be described as mountainous, and the descriptions of many of these craters in Chapters 4 to 7 of this book give the heights of crater rims as though they were ordinary mountains. In fact, most of the Moon's mountains are the result of impact mechanics. This is true even for the central peaks of lunar craters, formed when shock

Figure 2.12. The Alpine Valley. *CCD image by Maurizio Di Sciullo.*

waves from the initial impact travel upwards from the crater floor, which rebounds to construct piles of rubble thousands of meters high.

When reading descriptions of lunar features, you will frequently come across the rather vague term "hill," which can mean one of several things. Hills on the Moon can be just low mountains or modest central peaks, or they can be randomly distributed rubble piles within a crater. Sometimes the collapsed interior walls of large craters, the result of a process known as "slumping," forms hilly regions toward the edges of the crater's floor. The term "hill" may also be used to describe lunar domes.

Looking like lunar hills but frequently capped by tiny craterlets suggesting a similarity to terrestrial shield volcanoes, *domes* are the best evidence we have of volcanic activity on the Moon. Because they are small – on average only 5–10 km (3–6 miles) wide at their base – and fairly flattened – only 30–50 m (100–150 ft) tall – many lunar domes are difficult for the amateur lunar observer to detect, but some, like the feature designated Kies π and those around the

crater Marius, on the shores of Oceanus Procellarum, are easy to find with typical amateur telescopes. I can make out the crater pits atop many of the Marius domes in my 250 mm (10-inch) telescope, and they are easy objects in my friend's 405 mm (16-inch) reflector. In either case, you will need low Sun angles to observe lunar domes, because of their low profiles. Some lunar domes even display evidence that lava from these summit craters has flowed down their sides, resulting in little rilles. Others lack summit craters altogether, and may represent the equivalent of terrestrial cinder cones.

Now that we've covered the different types of lunar topography that you are likely to encounter in your telescopic explorations of the Moon, let's turn to the fundamentals of lunar geology, to learn how these topographic features formed in the first place. This exercise, which I call "A Crash Course in Lunar Geology," will hopefully give you some insights into the Moon's geologic history, though it is guaranteed to raise as many questions as it answers.

Chapter 3
A Crash Course in Lunar Geology

While recently perusing an excellent Website devoted to the Moon, I read a rather scathing review of a lunar observer's handbook published about a decade ago. I was a little surprised by the low opinion the reviewer had for this book, as I had enjoyed it myself – it was written by a longtime amateur lunar observer who I thought had done a nice job of conveying the enjoyment you can get from spending time at the eyepiece studying the wealth of detail our natural satellite has to offer. The reviewer's main complaint was that this handbook ignored our current knowledge of lunar geology, which has advanced by leaps and bounds since World War II, accelerated by the convoy of American and Soviet spacecraft sent to the Moon in the 1960s and early 1970s which climaxed in Project Apollo, the only mission to successfully return to Earth significant quantities of another world's rocks. More recently, the Clementine and Lunar Prospector missions have infused new energy into lunar studies, particularly its geologic aspects.

There once was a time when the Moon and planets were studied by scientists who took their degrees in astronomy. Through the early years of the twentieth century, these astronomers attempted to deduce the composition and climate of our Solar System's alien worlds by directly observing them through large telescopes, usually refractors (a telescope that uses an objective lens made from two or three glass elements), and sketching what they saw. Since the latter part of the nineteenth century, photography had been used to supplement the visual observations, but early attempts to photograph the Moon and planets were disappointing, and often showed much less detail than the eye could detect. Indeed, if you ever have the opportunity to examine the early photographic lunar atlas by Maurice Loewy and Pierre Henri Puiseux, or that by William H. Pickering, you might be surprised by how grainy and blurry the images are – certainly far less sharp and focused than the view through your own small telescope is likely to be!

Lunar Geology – What Have We Learned?

The astronomers who studied the Moon and planets through the first part of the twentieth century usually had good, solid backgrounds in celestial mechanics and could very adequately describe phenomena like the complex motions of these bodies. In fact, excellent work on the so-called lunar theory, the study of the Moon's very complex orbit, was done during this period of astronomical history, culminating in the work of Ernest W. Brown (1866–1938) and Dirk Brouwer (1902–66) at Yale University. But these astronomers had little or no training in geology or geophysics, nor in sub-specialties like atmospheric science. So they were ill-prepared to make sense of the topography they were observing on the Moon and elsewhere in the Solar System, and the result was sometimes a wild theory like Percival Lowell's Martian canals.

Today, all of that has changed. If you were to attend one of the annual meetings of the American Astronomical Society's Division for Planetary Sciences (DPS) you would discover that most of the scientists who study the Moon and planets today are geologists and geophysicists. Likewise, one of the leading scientific journals on solar system studies is published by the American Geophysical Union (AGU). Instead of looking through telescopes and drawing what they see, planetary scientists today perform most of their work on data transmitted by spacecraft sent to their targets, or in the laboratory analyzing specimens of lunar rocks and soil and meteorites, or perhaps on powerful computers that are able to model the geology and climates of other worlds. Simply put, if you don't understand geology, you are not likely to understand the work of today's professional planetary scientists.

Of course, it is hardly necessary for the amateur lunar enthusiast to procure a degree in geology in order to understand or appreciate the Moon, though the visual lunar observer is a kind of throwback to the old-time astronomers who peered at their quarry through the eyepiece. There is nothing wrong with that, and most amateur lunar observers I know have no interest in becoming experts in lunar geology or mineralogy. There can be no doubt that there is pure pleasure in just gazing through your telescope at the majesty of the lunar landscape. But it is not too difficult to learn a little bit about our Moon's geology and geologic history, and the amateur lunar observer who takes the time to do this will find that they will better understand and appreciate the vistas that unfold in the eyepiece.

The previous chapter introduced to the fundamental topographic features of the Moon, but made little or no attempt to describe how they got to be the way they are or what their geologic significance is. In this chapter I shall try to do those things, and more. I shall also discuss the way in which the Moon was probably formed, and spend a fair amount of time looking at questions about the Moon's geology that have yet to be answered. This section of the chapter may surprise you, for many people (incorrectly!) believe that the Apollo missions of 1967–72 told us everything we need to know about the Moon. Later in the book (Chapters 9–12) we shall see that there is still room for the amateur astronomer to contribute to our knowledge of the Moon, and we shall look at some very worthwhile projects being carried out by advanced amateurs.

It should not surprise you to learn that for a long time lunar geologic studies were obsessed with explaining the nature and origin of the lunar craters. After all, craters dominate the lunar landscape. They are the main feature that distinguishes the Moon from the Earth, where well-preserved craters are comparatively rare. If you are lucky enough to live near a library well-stocked with older books on the Moon, you will quickly discover that until roughly World War II, many who studied the Moon's craters were convinced they were looking at volcanoes similar to those that erupt on our planet.

As we saw in the previous chapter, it was not until the work of people such as Ralph Baldwin and Gene Shoemaker, undertaken in the second half of the twentieth century, that we began to slowly realize that impacts of asteroids and smaller meteoroids, not volcanoes, were the foremost forces responsible for shaping the Moon's topography. And while we have lunar lava flows to thank for the smooth, dark maria, we now understand that this kind of volcanism was quite different from the volcanic activity we observe today on Earth.

While many persons mistakenly believe that the US–Soviet "space race," from the late 1950s to the early 1970s, merely satisfied political objectives for the two competing countries, nothing could be further from the truth. Not only was sending the many unmanned spacecraft and the manned Apollo missions to the Moon a major technological achievement for all mankind, but these missions successfully completed a mind-boggling array of scientific experiments. From the mid-1960s to the mid-1970s, five Lunar Orbiter spacecraft and eight Apollos photographed the Moon in unprecedented detail, specimens of lunar soil and rock were analyzed in situ by five Lunar Surveyor and two Soviet Lunokhod landers, six Apollo missions successfully returned a total of 382 kg of Moon rocks from six landing sites, and smaller amounts of lunar samples were returned by three unmanned Soviet probes.

Thanks to these explorations, we now have a far better knowledge of the composition of discrete lunar features and how they were formed, which in turn allows us to grasp the Moon's geologic history, and make educated guesses about how the Moon itself came to be. These are huge advances, yet we still have much to learn, for our newly acquired knowledge allows us to ask far more sophisticated questions, some of which would not have occurred to us before these new discoveries were made.

A Lunar Geologist's Lexicon

To understand the Moon's geology, we need to add to our vocabulary a few terms that will come up again and again. For example, it is helpful to understand the different basic types of rock that geologists have identified here on Earth – a classification scheme that is useful for Moon rocks, too. There are three basic categories of rocks – igneous, sedimentary and metamorphic, depending on how the rock formed. We can classify any rock from the Earth or Moon as one of these three basic types. *Igneous rocks*, which you may think of as cooled lavas, directly crystallize from silicate (minerals containing silicon and oxygen) melts. Silicate melts can produce a bewildering variety of minerals, depending on their initial composition before the cooling process begins, and to a far lesser degree, the rate of cooling. *Sedimentary rocks* are built up over time by the deposition of sediments, usually in a watery environment, which glues together the detritus of a wide variety of component rocks. The Moon lacks the typical sedimentary rocks, like sandstone, we find here on Earth, perhaps because it lacks water. However, lunar geologists consider the lunar *regolith*, the equivalent of our soil, to be a kind of sedimentary rock because of the way it forms. *Metamorphic rocks*, the third fundamental type, form from igneous or sedimentary rocks which are subsequently subjected to the transforming forces of high temperatures and pressures. The lunar breccias, described a little later in this chapter, are a good example of rock subjected to metamorphosing forces.

Among the most important families of lunar rock are the *basalts*. They are fine-grained, igneous rocks that make up the crusts of our own ocean beds and which are associated with volcanic processes. Figure 3.1 shows a typical lunar basalt. All basalts contain the same two minerals – pyroxene and plagioclase. *Pyroxene* is an iron–magnesium silicate mineral that exists in two basic forms – a "clino-" form that also contains calcium, and an "ortho-" form, which is the kind commonly found on the Moon. *Plagioclase* is a calcium feldspar mineral, also common on the Moon.

Basalts are associated with flows of thin sheets of lava, and we now are quite certain this is how the lunar maria formed. During the 1960s, the Surveyor spacecraft analyzed mare rocks and found them to be basaltic, so we had confirmation of the volcanic origin of the maria before Apollo visited the Moon. Apollo collected basaltic rocks from several mare regions – Oceanus Procellarum, Mare Fecunditatis, Mare Crisium, Mare Imbrium, and Mare Serenitatis. In fact, the first lunar landing mission – Apollo 11 in 1969 – returned nothing but basalts. In places the mare basins are filled to a depth of less than 1 km (about $\frac{1}{2}$ mile) but in other places the lava is much deeper – 4 km ($2\frac{1}{2}$ miles) or more.

Geologists working with the specimens returned to Earth by the Apollo missions were able to distinguish three sub-families of lunar basalts: titanium-rich basalts discovered by Apollos 11 and 17, titanium-poor basalts found by Apollos 12 and 15, and aluminous basalts (these were actually discovered by the Russian probe, Luna 16, which returned small samples of lunar soil). It was soon realized that lunar basalts differ from those on Earth, particularly in their lack of the so-called *volatile* elements (compounds that boil at low temperatures). Earth basalts contain far more potassium and zinc, for example, than their lunar counterparts, which also lack water.

The question of whether water is locked into the Moon's rocks or trapped at its frigid poles has long been of great interest. It is not unreasonable to assume that some of the impacting bodies to strike the Moon over the eons were comets containing large amounts of water ice. During its main mapping mission, NASA's Lunar Prospector probe found strong spectral signatures for hydrogen in the permanently shadowed craters at the Moon's south pole, so on July 31, 1999 the space agency deliberately crashed Lunar Prospector into one of these craters, hoping to kick up a cloud of water vapor that could be detected from Earth. The results of the experiment were negative – lunar water remains elusive, at least for the time being.

Lunar basalts have other interesting characteristics. For example, they are rich in iron, which gives them their dark color. You can see this color with the naked eye. They also have high levels of *refractory* elements (compounds that boil at high temperatures) such as titanium, zirconium, uranium, and the so-called rare earth elements such as lanthanum. The low ratios of potassium (a volatile element) to lanthanum which we find in all mare basalts shows that the Moon is depleted in volatiles. My favorite eyepieces for observing the Moon incorporate the rare earth lanthanum to give them excellent color correction – how appropriate it is to use such eyepieces to view the lanthanum-rich Moon!

Many lunar basalts also have a weird, porous appearance that resulted from many gas bubbles

Figure 3.1. A representative lunar basalt. The scale is in centimeters. *Courtesy LPI, Houston, Texas.*

being trapped within them as they solidified. They look as though they are the product of lavas that vigorously frothed as they erupted, which is highly possible in the vacuum of the lunar environment. Some of the lunar samples returned by Apollo 17, the last manned mission to the Moon, were classified as *quenched basalts*, a colorful name that indicates these rocks cooled very rapidly from the erupting lava, resulting in very fine-grained or even glassy-structured textures. Quenched basalts probably formed at the surface of lava flows where heat could be quickly and easily dissipated. The chemical composition of quenched basalts is probably not too different from that of the original lunar lavas that produced them, because their rapid cooling left little opportunity for changes to take place. Most lunar basalts, however, are very coarse-grained, suggesting a much more leisurely cooling process. Also very coarse-grained are the rocks found in the lunar highlands, to which we now turn our attention.

Besides the lava-smoothed mare areas, the Moon's other dominant topographic element is known as the lunar *highlands*, a rather generic term that encompasses the huge chunks of crust thrown up by impacts that ring the mare basins, the various mountains ranges, and the rims of large craters. In a pair of binoculars, the bright white lunar highlands contrast nicely with the dark maria. The lunar highland rocks are light-colored because they are rich in calcium, magnesium, and aluminum, rather than iron. What kinds of rocks do we find in the highlands, and what do they tell us about the Moon?

Interestingly, the first clues to the composition of the lunar highlands were found, not in the highlands themselves, but in the lunar regolith, or lunar soil, sampled by Apollo 11. As well as mare basalts, these soil samples contained pieces of *breccias*, rocks made from a matrix of stony fragments and a finer component that binds the fragments together. Granite is an important component of some of these breccias. Later missions would find an abundance of breccias in the lunar highlands. Figure 3.2 shows a typical lunar breccia, colorfully nicknamed "Big Bertha."

But the real treasure found in the lunar regolith turned out to be two rather exotic minerals that would gain new importance during the heyday of Apollo – anorthosite and KREEP. *Anorthosite* is a coarse-grained rock, largely composed of calcium feldspar. We now believe that anorthosite is the dominant constituent of the lunar crust. The acronym *KREEP* stands for potassium (whose chemical symbol is K), rare earth elements, and phosphorus, chosen to reflect the rich abundance of

Figure 3.2. A typical lunar breccia, nicknamed Big Bertha. The scale is in centimeters. *Courtesy LPI, Houston, Texas.*

these elements in this mysterious mineral. Besides anorthosite, KREEP, and granite, a fourth rock type – called *anorthositic gabbro* (gabbro is simply a course, crystalline rock consisting of plagioclase and pyroxene) – was found by later Apollo missions to be an important component of the lunar highlands. Anorthositic gabbro was regarded as the elusive "highland basalt." Lunar geologists are intensely interested in finding and studying basalts in the highlands, because presumably these basalts are quite ancient, left over from volcanism that predated the mare-filling volcanism.

One of the more fascinating aspects of the lunar highland rocks is that they show evidence of what geologists term *shock metamorphism*. This simply means that the original structure of the rock has been transformed by a great force, such as the impact of an asteroid that created one of the great mare basins. We must keep in mind at all times that the lunar highlands are just fragments of the Moon's crust uplifted and re-arranged by impacts. The coarse-grained highland rocks show so-called cataclastic textures which form by the violent crushing of crystals. These textures are characterized by a fine matrix of crushed rock and mineral fragments surrounding larger fragments of the same minerals. Many anorthosites show cataclastic structures. Further evidence for impact shock was discovered in recrystallized breccias, including one collected by the Apollo 15 astronauts that was dubbed Genesis Rock, and in the glassy breccias from the Apennines Front – the steep, cliff-like western face of the Apennines – also collected by Apollo 15.

Our Changing Views of Lunar Geology

Surprisingly, despite all of this mineralogical treasure and the discoveries that followed from its study, few people, including most amateur lunar observers I know, are fully aware of how profoundly the space missions of the 1960s and early 1970s affected our view of the Moon's geology and geologic history. Perhaps this is because the science behind these discoveries is complicated, or perhaps neither the governments that sponsored these missions nor the scientific community who were involved with them bothered to try communicating these exciting discoveries to the public in a comprehensible way. Before we go any further, let's summarize how our space-age exploration of the Moon, particularly Project Apollo, transformed our view of lunar geology.

Let's start with the lunar craters, since they are the most important distinguishing feature of the Moon and we've already discussed them at some length. Before the lunar spacecraft missions, many reputable scientists thought they were just lunar volcanoes, despite much research suggesting that impacts of asteroids and meteoroids were the dominant cratering mechanism. In 1893, for example, the geologist Grove Gilbert (one of the first geologists, rather than astronomers, to turn their attention toward the Moon) published a very important paper in which he correctly deduced that lunar craters were not volcanic, but that the mare regions were indeed the product of vast sheet flows of lava.

For a long time Gilbert's work was forgotten, until the 1940s when Ralph Baldwin independently arrived at the conclusion that most lunar craters were impact sites rather than volcanoes. Nevertheless, there were many who remained proponents of the lunar volcanism theory of crater formation until space missions like Lunar Orbiter obtained unequivocal images showing that many structures suspected of being volcanoes were not, while at the same time confirming what scientists who studied the forces of impacts on Earth had predicted for lunar impact sites. Apollo determined the dynamics of debris ejected by impacts, removing one of the last objections to the impact theory.

Apollo also proved that the Moon was very dry – the Moon rocks and soil had no water, and in general were devoid of volatile compounds altogether. Before Apollo, some planetary scientists were still speculating that water had once flowed over the lunar surface, in places like Schröter's

Valley. Now we know that it is very unlikely that water flowed over the Moon's surface, and it is virtually certain that water was never a major force in shaping the lunar landscape. Schröter's Valley is a lava tube, not a dried-up river bed. These findings have important implications for theories of how the Moon may have formed.

One thing now appears very clear – the Moon was subjected to extreme heating during its genesis. Very high temperatures are required to drive off water and other volatile compounds – temperatures that could have resulted from a collision between Earth and some unknown, ancient impactor. This is just one reason for the current popularity of the *giant impact theory*, first proposed in the 1940s by the Canadian geologist Reginald Daly of Harvard University, to explain the Moon's origin; we shall discuss other reasons for its acceptance later in this chapter. This theory proposes that the Earth was struck by a Mars-sized planetoid 4.6 billion years ago, the heavier parts of the impactor sticking to the Earth, while the lighter parts rebounded into Earth orbit. There they gradually coalesced, over the next 60 to 80 million years, with a mixture of generous amounts of the Earth's crust also thrown into space by the impact and with debris from the young Solar System that continued to accrete long after the impact.

The return of lunar rocks and soil samples by Apollo allowed scientists to positively date those rocks, showing that the Moon was as ancient as the Earth. The highland rocks, remnants of the lunar crust, showed ages of 4.1 billion years or older, with some anorthosites as old as 4.4 billion years. The lava basalts varied between 4.3 and 2 billion years, which gave geologists important clues about the age and duration of lunar lava flows. Before Apollo, geologists had no way of really knowing what the highland or mare rocks were made of, but we learned that basalts dominated the mare lavas and that the highlands were comprised of a complex mixture of rock types, all having much more aluminum than the basalts.

Thanks to Apollo, we were also able to learn something of the mantle, the layer of fluidized rock lying below the lunar crust, and we discovered that this layer is likely composed of olivine and pyroxene. This gave rise to the "magma ocean theory," according to which an ocean of magma some 400 km (250 miles) deep is thought to have once encircled the Moon's solid core, slowly crystallizing over a period of 100 million years. According to this theory, lighter materials like the plagioclase feldspars (anorthosite) would have risen to the surface to form the lunar crust, which is why we find so much of these minerals in the present-day lunar highlands. Today the lunar crust is about 60 km (35 miles) thick on the Moon's nearside, 100 km (60 miles) thick on the farside. Heavier minerals like olivine and pyroxene would have sunk below the surface to form the mantle. The magma ocean would have been hot enough to boil off the volatile elements that are so noticeably absent from the Moon today.

It is also highly significant that virtually all of the lunar rocks collected by Apollo showed identical oxygen isotopic abundances to terrestrial rocks, which means that the Earth and Moon had to form in the same neighborhood of the Solar System. This finding also defeated the "capture hypothesis" which attempted to explain the Earth–Moon system by invoking a very unlikely trick of celestial mechanics: that the Earth managed to grab onto a Moon that was somehow thrown into the inner Solar System from elsewhere. If that hypothesis were true, the Earth and Moon would not have rocks with identical oxygen isotopes.

How the Moon Came to Be

Apollo also allowed planetary scientists to reject the "fission hypothesis" of Moon formation, according to which a very rapidly spinning young Earth would have developed such an extreme bulge at its equator that it would have been compelled to "spin off" a satellite. No one could figure out how to get the young Earth spinning fast enough (every 2.5 h!) to do this, nor could they explain why today's Earth–Moon double planet has nowhere near the momentum necessary to initiate spin-off, or alternatively where the missing rotational energy went. But the final blow to the fission hypothesis came from Apollo's finding that, aside from the identical oxygen isotopes, the basic compositions of Moon rocks are very different from those found on our planet. If the Moon had spun off from the Earth it should be made from exactly the same stuff, that is, from the same rocks we find in the Earth's crust and mantle (the Earth's iron core would presumably have not been involved in the spin-off).

But as we have seen, such is not the case. Apollo found that lunar rocks lack volatile compounds like water, potassium, and sodium, which are all found in great amounts in Earth rocks. But lunar rocks *do* have abundant refractory compounds – aluminum, calcium, thorium, and the rare earths, up to 50%

more than typical Earth rocks do. The ratio of iron oxide to magnesium oxide is also different, by about 10%. These findings also spelled trouble for the third classical theory of lunar genesis – the "double planet hypothesis," which argued that the Earth and Moon formed from the same embryo of gas and dust in the early Solar System.

If this theory worked, the two should have basically the same composition, which we know is not true. There are other problems with this theory as well – it fails to explain why Earth has a massive metallic core while the Moon's is tiny, or how the Earth's day came to be 24 h long, which is faster than predicted by the double planet model, which would only allow small increments in angular momentum from the accretion of debris orbiting the nascent Earth.

The giant impact theory solves many of the problems with the classical theories while explaining most of Apollo's findings. This theory was proposed in the 1940s but not revived until the 1970s by several different planetary scientists – William K. Hartmann, Donald R. Davis, A.G.W. Cameron and William R. Ward. One of the problems with the capture theory was that a planetoid that closely approaches Earth is much more likely, by the laws of celestial mechanics, either to strike the Earth or to veer past it in a "slingshot" (receiving a gravitational boost by its close approach) than to be captured in a neat, orderly orbit around Earth. The giant impact theory calls for a collision between the two bodies, which the odds favor. But the theory had more going for it than this after Apollo.

For example, it explains why the Earth has a large metallic iron core but the Moon does not – if the impactor had an iron core of its own, it would have coalesced with the Earth's core, the Moon forming from the lighter parts of both objects – the silicates – which is just what Apollo shows is the case. If the Moon formed primarily from the impactor, with a little bit of Earth crust thrown in for good measure, this would explain the 10% discrepancy in the iron oxide/magnesium oxide ratio that we in fact see. The giant impact theory also explains the differences in the volatile and refractory abundances of Earth and Moon rocks – the collision would have produced very high temperatures that would have boiled away the Moon's volatile elements, but its refractory elements would have better survived the high temperatures, recondensing quickly to form a disproportionate component of the lunar crust.

The giant impact theory also satisfies scientists who study the physics of the Earth–Moon system because it explains how the Earth's day is so short. Without such a collision, the Earth should spin on its axis much more slowly – if, that is, we correctly understand how planets form from the disk of gas and dust that presumably encircled our Sun early in the history of the Solar System's evolution. Astronomers have discovered such disks of primordial planet-forming material around other stars in our Galaxy, and we may someday have the opportunity to watch new planets being born before our very eyes.

The problem has always been how to spin up the Earth to its present-day rotational rate. The impact of small meteorites won't do this – for one thing, their aggregate effect is too small; for another, their effects on angular momentum tend to cancel out, because some will strike the Earth in such a way as to speed it up in one direction, while others will strike our planet from the opposite direction, having the opposite effect. But one giant impact, imperfectly centered, would tend to impart a tremendous, one-time spin-up to the Earth's rotation that would act in one direction only. Calculations suggest that the impactor would have been about the same size as the planet Mars is today. For all of these reasons, the giant impact theory is widely accepted today as the best explanation of the Moon's origin.

The Lunar Geologic Timescale

But Apollo and the other lunar missions have also told us much about the geologic history of the Moon following its birth. Lunar geologists have adopted a timetable of lunar geologic history similar to the better-known timeline "eras" in the Earth's geologic history. As we have seen, the Moon was likely formed by the impact of a sizable planetoid with our own Earth some 4.6 billion years ago. Lunar scientists have designated the earliest period in lunar geologic history as the *Pre-Nectarian Era*, so-called because it describes the condition of the Moon from its formation up to the much smaller (but still sizable as far as the Moon was concerned!) impact that created the Mare Nectaris basin. The Moon underwent its most violent transformations during this period of its geologic history.

If the magma ocean theory is correct, it is quite likely that this ocean may have been a consequence of the original giant impact with Earth, for calculations suggest that such an impact could have melted up to a third of both impactor and Earth. If the original impact did not provide the heat energy to form the magma ocean, or if the effect wore off as the remains of the impactor re-formed and cooled in Earth orbit, there are other mechanisms that could account for the magma ocean. For example, the accretion of primordial Solar System debris by the very young Moon was not a gentle process, but would have involved very energetic impacts capable of melting the Moon's outer layers to create the magma ocean. The heat released by sinking metallic iron during core formation could have provided the melting energy, as could tidal interactions with the Earth. The heat released by the decay of radioactive lunar isotopes may have contributed to this melting process, or may have been the sole cause – Apollo discovered huge amounts of radioactive uranium, thorium and potassium isotopes in the lunar crust that would have released huge amounts of heat, assuming they existed in similar concentrations throughout the young Moon's crust and mantle.

Eventually the magma ocean would have cooled, but not before boiling off volatiles like water, sodium and potassium. Low-density plagioclase feldspars (anorthosite) rose to the top of the ocean, forming the solid crust. Below the light-colored crust of anorthosite, the magma would have continued to cool and solidify down to the depth of 1100 km (700 miles) determined by Apollo's seismic wave experiments as the location of the boundary between the solid and molten zones. This discovery shows that the Moon is mostly solid, since only the inner core, of radius 600 km (375-mile), remains molten today. Moonquakes still occur in the mantle at depths of 600 to 800 km (375–500 miles) below the surface, so the Moon is not completely quiet. There is evidence that tidal interactions with Earth trigger many of these moonquakes, which tend to occur when the Moon is at perigee or apogee. The magma ocean probably solidified about 4.2 billion years ago.

As you might guess, the *Nectarian Era* followed the Pre-Nectarian Era, beginning with the great impact event that created Mare Nectaris, a beautiful dark oval easily visible in binoculars in the Moon's southeast quadrant. This event occurred 3.92 billion years ago, and was just one of numerous significant impacts by meteoroids suffered by the Moon during this period, creating a multi-ringed impact basin whose outer ring is partly preserved as the Altai Scarp (see Fig. 3.3). The most significant happening of the Nectarian Era is what planetary scientists have colorfully called the "late heavy bombardment" or "terminal cataclysm." During this period, which ended 3.9 billion years ago, small asteroids blasted out more than forty large impact basins from the young lunar crust. How do we know this?

Figure 3.3. The outer ring of a multi-ringed impact basin, partly preserved as the Altai Scarp. *Video image by Joe LaVigne.*

The Apollo missions sampled highland (crustal) rocks from widely separated sites. Despite being from locales that were very distant from one another, isotope dating of these rocks showed ages that clustered around 3.9 billion years. This suggested that most of the lunar highlands formed 3.9 billion years ago. This was a surprise to lunar geologists, who had assumed that the ages would be closer to 4.2 billion years, when the magma ocean solidified. Clearly the basin-forming impacts had reset the geologic clocks of these highlands crustal rocks to 3.9 billion years. These rocks, including the breccias, all showed signs of impact shock, as we have seen. By contrast, Apollo found that rocks younger than this are better-preserved, with few shock effects. The heavy bombardment of the lunar surface during the Nectarian Era was also responsible for creating the 2 km (1.2-mile) deep regolith layer by pulverizing the crust.

The next era of lunar geologic history, following the Nectarian Era, is called the *Imbrium Era* because it begins with the giant impact that formed Mare Imbrium 3.85 billion years ago, not long after the end of the late heavy bombardment. The impactor that formed Imbrium was probably an asteroid some 100 km (60 miles) in diameter, its collision with the Moon causing a catastrophic explosion that carved out a huge crater where there is now a circular dark area easily visible to the naked eye. Mare Imbrium dominates the northwest quadrant of the Moon, measuring 1500 km (nearly 1000 miles) across its lava plains.

The Imbrium impact would have shaken the Moon so badly that it likely touched off a vigorous period of lunar volcanism – numerous fractures in the thin, young lunar crust would have allowed huge amounts of lavas to well up from the molten zone, which was not nearly so far below the surface at this period in the Moon's geologic history. Over the next 700 million years, these lavas filled the forty-plus impact basins carved out during the preceding Nectarian Era, which is why we see these areas as dark-colored today. You can see a lot of the Moon's geologic history just by looking at Fig. 3.4, which is an unusual mosaic of near-infrared photographs. Because infrared (IR) wavelengths are slightly longer than the wavelengths of visible light, in photographs such as this the differences in the composition of lunar features is slightly exaggerated. The basalts that make up the maria absorb IR radiation (heat) very readily, whereas the brighter brecciated regions tend to reflect IR. As a result, the very dark maria contrasted against the much lighter highlands, and the ray systems associated with the freshest impact sites.

It should be explained that the lunar lavas were far different from the lavas produced by terrestrial volcanoes. Samples of the lunar basalts collected by Apollo astronauts showed that the lavas were less viscous than their Earthly counterparts. Laboratory experiments showed that lunar lava has the consistency of thin motor oil, not thick enough to build the shield volcanoes we find on our planet. These thin lavas would have had a tendency to pond at the terminus of the eruption, covering many of the lava tubes and other telltale signs of volcanism that are strangely rare on the Moon. But they were dense enough to stress the light, underlying plagioclase crust, causing it to fracture and subside, forming the concentric rilles you can see today around the margins of many mare regions, including Mare Humorum. It is possible that the mysterious wrinkle ridges, or at least some of them, were also produced in this way.

We might also ask why there is an absence of lava-filed mare regions on the Moon's farside. The best answer to this perplexing question is probably that there was no Imbrium-scale impact on the farside when the lunar crust was thin enough to permit lava upwellings. By the end of the Imbrium Era, 3.15 billion years ago, the lunar crust had solidified to a depth where the lavas were too deep to make it to the surface, terminating the period of widespread near-side volcanism. By this time the Moon looked much like it does today, with all the major mare and high-lands regions we recognize already formed.

Following the cessation of lunar volcanism, we next come to the longest period in lunar geologic history, the *Eratosthenian Era*, lasting from about 3 billion years ago to 1.2 billion years ago. There were no cataclysmic, Imbrium-style impacts during this

Figure 3.4. An unusual mosaic of near-infrared photographs contrasting the dark mare regions with the much lighter highlands. *Photographic mosaic by Eric Douglass.*

era, nor was there the widespread volcanism that filled the mare basins. Instead, the Moon experienced a little bit of volcanism during this time, and many modest impacts of meteoroids that, over 2 billion years, gradually eroded the highland crust and pulverized the surface, softening and rounding the appearance of not only the lunar mountains, but of just about every other aspect of the Moon's topography. Surely many formations were completely destroyed by the eons of meteorite impacts, and it

is likely that much evidence of multiple impact rings was either obliterated or rendered obscure by this process. Figure 3.5 shows the Eratosthenes/Copernicus region, which includes the craters named for this period of lunar geologic history and the next.

The most recent period of lunar geologic history is termed the *Copernican Era*, because the great crater Copernicus, which is 810 million years old, formed not too long after the beginning of this era. Beginning about 1.2 billion years ago, the

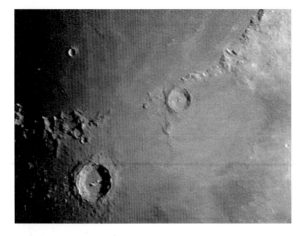

Figure 3.5. The Eratosthenes/Copernicus region. *Video image by Joe LaVigne.*

Copernican Era is characterized by a few moderate impact events that created relatively fresh craters like Copernicus and Tycho, which is "only" 109 million years old. Figure 3.6 is a geologic map of Tycho. Copernicus's and Tycho's well-preserved ray systems are evidence of their relative youth, as far as lunar features go.

Lunar Geology's Unsolved Mysteries

So far we have concentrated our discussion in this chapter to what we know about the Moon's geology and its probable origin and geologic history, and it is clear that, thanks to Apollo and other spacecraft missions, we have been able to answer many questions on these topics. But are there important questions left unresolved? The answer, of course, is that we have only just begun to understand lunar geology. In fact, it is safe to say that the greatest legacy of all the lunar missions is that now, for the first time, we are finally able to start asking truly meaningful questions about the Moon's origin and geologic processes. The rest of this chapter will pose a few of these questions – perhaps readers of this book will someday answer some of them!

Let us start with the Moon's birth. Although the giant impact theory is now accepted as the best explanation to date for the varied compositions of the Earth and its Moon, the evidence is far from conclusive. For example, we would like to find some unequivocal geochemical signature in the lunar rocks for both the impactor and the Earth – perhaps we shall discover that signature during a return visit to the Moon, a visit that is badly needed since we have explored only a tiny fraction of its interesting surface. And how has the Earth–Moon system dynamically evolved since the Moon's birth, assuming the that impact-ejection hypothesis is correct?

It seems pretty certain that massive collisions were the order of the day in the early Solar System, 4.6 billion years ago, and that the rate of these collisions has steadily declined since that time, when the Moon would have accumulated mass by accreting primordial matter. But how sure can we be that there was in fact a "late heavy bombardment" 3.9 billion years ago? What if we found many lunar rocks older than that during future missions? What if debris ejected from the Imbrium impact has polluted the lunar rock-age data to obscure evidence of older rocks? Could the age clustering around 3.9 billion years represent the final throes of primordial accretion instead of a separate bombardment? Were the ages of lunar rocks continuously reset by the latest catastrophic impacts? And if the answer to this last question is "yes,", how did Apollo 14 find lavas that were 4.3 billion years old? If a late bombardment did occur, what exactly happened from the time of primordial accretion to the end of this bombardment, and are there any telltale signs of this transition? Has the rate of cratering remained constant during the past three billion years?

And what of the magma ocean theory? Can we ever narrow down the many possible mechanisms for its formation? And does it really account for the differentiation of minerals we see on the lunar surface? Why do some highland crustal rocks lack the feldspars that should be ubiquitous if the magma ocean theory is right? What kinds of rock were formed in the magma ocean? We still don't know enough about how minerals form in magma, or how the partitioning of trace elements occurs between the magma and the crystallized end products created by a hypothetical magma ocean, to answer these questions. If the magma ocean theory is not correct, how did the lunar crust form?

Whatever the mechanism for the formation of the lunar crust, many questions have gone unanswered about the specifics of the crust itself. How many different types of magma contributed to the formation of the lunar crust? When did these magmas become mixed into breccias? What exactly are the global variations in the thickness of the lunar crust, and what is the cause of these variations? Why is there a discrepancy between the Moon's center of mass (where its density "balances out") and its center of figure (its geometric center)? Did the

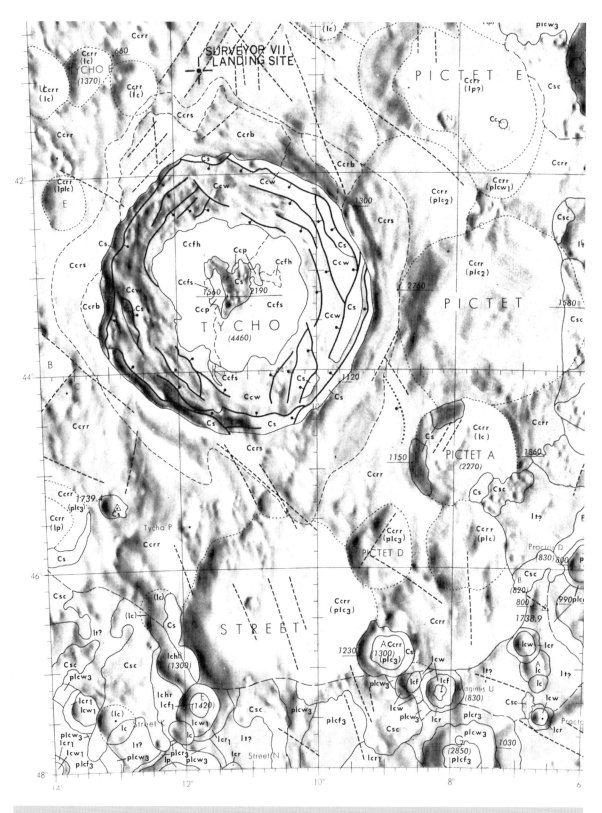

Figure 3.6. A geologic map of Tycho. *Courtesy USGS, Flagstaff, Arizona.*

mighty impacts that formed the largest maria exca-vate complete sections of the primordial lunar crust that can be reconstructed by studying ejecta? What mechanisms of isostasy predominate in the lunar crust today? Are there layers to the lunar crust? These are all difficult questions to answer.

Nor do we know with any degree of certainty the precise ages of all the mare basins. For example, we believe that the Mare Serenitatis impact occurred first, closely followed by the impacts that created Mares Nectaris, Crisium and Humorum at perhaps 3.9 billion years, then the Imbrium impact 3.85 billion years ago. Accurately dating the various basins will require samples of impact melts from each of these basins, but we don't have such speci-mens. Closely related are questions about the dura-tion and nature of lunar volcanism. We believe that filling of the eastern maria (Tranquillitatis, Crisium, and Fecunditatis) took place from 3.85 to 3.3 billion years ago, followed by the formation of the western maria (Imbrium and Procellarum) from 3.3 to 3 billion years ago, when extensive lunar volcanism appears to have ceased. It would be interesting to know how old the ancient highlands basalts are and where they are exposed; conversely, what are the ages of the youngest mare basalts?

As for lunar craters, we still aren't sure exactly how their central peaks or rims form – we still depend heavily on theoretical models of impact dynamics to explain these structures. How does the velocity of the impactor affect the morphology of the crater it creates, and what are the precise mech-anisms for making crater chains? What minerals were contributed to lunar rocks by the impactors? Are impact melts chemically homogenous? How do fractures on the floors of craters form? And what of

the multi-ringed impact basins like Mare Orientale (see Fig. 2.7) – did these all form at about the same time, and exactly how energetic were the impacts required to create these impressive structures?

The list of unanswered questions is almost endless. What was the nature of lunar highland vol-canism, and when did it cease? How did the wrinkle ridges form? What other evidence is there for tec-tonic activity on the Moon? Are the concentric rilles at the mare perimeters the result of the basin-forming impact or of subsequent stresses by basin-filling lavas? What is the origin of the exotic KREEP? Did impacts or volcanism add KREEP to the lunar surface? What are the exact components of the lunar mantle? How did lava get from the mantle to the surface during the age of widespread volcanism? Can it still do so, or is it trapped beneath a deep, unyielding crustal layer? Does our Moon still contain any molten materials deep inside it? Results obtained by the Lunar Prospector in the late 1990s suggest that the Moon has a small iron core. What does this tell us about the origin of the Moon?

Keep these questions in mind as you begin to explore the Moon's most interesting topographic features, described in the next four chapters. Despite the need for better answers to the many questions surrounding the geology of the Moon, the Apollo landings and all the other lunar missions have rewarded us with keen insights into why the Moon looks as it does and how it got that way. Hopefully you have picked up at least a little under-standing of lunar geology by reading this chapter, so that as you survey the Moon's maria, bays, craters, rays, mountains, domes, rilles, and wrinkle ridges you will not only appreciate their great beauty, but also find them scientifically interesting.

Chapter 4
Lunar Features – Northeast Quadrant

This and the next three chapters are really nothing more than an illustrated selective catalog of some of the more interesting lunar features which the amateur astronomer equipped with a telescope as small as 75 mm (3 inches) can expect to observe. I have tried to include most of the more obvious lunar features, but this is not an exhaustive list of all the maria, walled plains, craters, mountains, rilles or other features that deserve the attention of the amateur lunar observer. Such lists were long ago compiled by the many authors of the classic lunar observers' guides, and several of these are identified in Further Reading at the end of the book, and also in the longer bibliography on the CD-ROM.

I have decided to keep alive the tradition of using terms such as "walled plain," "rampart," and "glacis" that are not generally preferred by today's professional lunar scientists. I have done so both because these terms are used in the old guidebooks, and because they are very descriptive and thus helpful to the amateur observer. The dimensions of lunar features are given in metric units, with equivalents in miles or feet. In the case of the elevations, the main sources of published values are the old guidebooks, where they are quoted in feet; the metric equivalents have been calculated with due regard for the accuracy of the original measurements.

The illustrations are hand drawings rather than photographs. They are all the work of amateur astronomers of considerable talent, and they show what amazing detail can be recorded by the careful observer who takes the time and trouble to make a permanent record of what they see. In Chapter 9 I shall cover the basics of how to make such drawings, so that you can try your own hand at this rewarding activity.

For the purposes of this survey the Moon is divided into four quadrants – northeast, southeast, northwest, and southwest. All of the features described in this book will be found on any good lunar map or atlas (many of these observing aids are also listed in the bibliography on the CD-ROM). The notes on each feature are mostly descriptive, to help the observer know what to look for, but in some instances I have also tried to include a few notes about the geologic events that might have created the features, so that the observer can better understand what he or she is looking at. So, with the preliminaries out of the way, let's start exploring the Moon!

Mare Crisium, the Sea of Crises, is a huge impact site 3.85 billion years old. It provides a good jumping-off point for finding lunar features in the northeast quadrant, as it is easily visible to the unaided eye as a darkened oval. Mare Crisium is an excellent example of the optical illusion of foreshortening, in which features near the limb appear to be squashed along their east–west axes. Although Mare Crisium is actually wider along its east–west axis than along its north–south axis, to our eyes just the opposite appears to be true, as much of the east–west distance is compacted by its proximity to the eastern limb of the Moon. Mare Crisium's relatively smooth lava-flooded surface measures about 500 km (300 miles) across and takes up 170 000 km^2 (66 000 square miles). Unlike some of the other maria, whose borders have been greatly eroded, Mare Crisium is well preserved, its mountainous borders still intact after all those eons. Near the southwest border of Mare Crisium, in the neighborhood of the Agarum Promontory (see below), these mountains rise to heights of 3500 m (11 000 ft).

The floor of Mare Crisium is almost featureless, save for a half dozen craters on the west side, and half that many just inside the northern rim. Why are there no craters in the middle of Crisium? In the first place, many if not all of the older craters that existed before the impact event that created Mare Crisium would have been destroyed by that impact. We can therefore be confident that any craters we see inside the maria are younger than the maria themselves. Still, you might ask, wouldn't some cratering be expected to have taken place after the main impact that created Mare Crisium, and if so, why don't we see more of these craters? The most likely explanation is that, since most impact basins like Mare Crisium are bowl-shaped – deeper in the middle than at the margins – any craters created after the main impact event in the deeper parts of the crater would now be concealed under the lava flows that have occurred since the impact event. Of the craters that formed after the main impact and before the lava flooding, only those at the edges of the mare, where the ground is highest and most unaffected by lava flows, were not obliterated. Of course, it is possible to find "fresh" craters sitting atop a solidified lava flow, but these tend to be few and far between, proving that most of the Moon's craters were created early in its geologic history.

Agarum Promontory is a bright cape that projects into Mare Crisium's southeast border. See if you can detect the two craters on its summit. Several mountain peaks are also visible here, the highest, in the east, rising some 3500 m (11 000 ft) above the plain of Mare Crisium. The east side of Agarum contains a wide bay with a high ridge on its west side, and there is evidence of several ruined crater rings in the vicinity. One of the more interesting features associated with Agarum is an extremely long wrinkle ridge that begins at Agarum and runs northward all along the eastern boundary of Mare Crisium. About halfway along this ridge is a group of bright mountain peaks just off to the east of the ridge, just west of the mountains bordering Crisium on that side of the sea. Several of the classic observers of the Moon have claimed to have seen "mists" in Agarum Promontory, at times when Mare Crisium is bisected by the terminator under a low angle of illumination at sunrise. This is an example of the controversial lunar transient phenomena (LTPs), discussed in Chapter 10.

Alhazen, close to the east edge of Mare Crisium, is a modest-looking crater, typical in all respects, with terraced walls and a central peak. Try to detect the ruined ring just south of this crater that connects it to Hansen. Easier to see is the low-walled ring plain Recorde, northeast of Alhazen.

Picard is the largest crater to be found on the relatively smooth floor of Mare Crisium, which isn't saying very much. It is only 34 km (21 miles) across,

but rather deep, with a rim that is 2500 m (8000 ft) above its floor.

Yerkes is what Hugh Percy Wilkins and Patrick Moore in their classic guide *The Moon, A Complete Description of the Surface of the Moon* called "the relic of a once-complete ring." It lies on the floor of Mare Crisium, just a little west of Picard. Named for the Chicago street-car tycoon whose generosity built the famous Wisconsin observatory of the same name, the east wall of this ring is almost completely invisible – this is what is meant by a ghost crater. The west wall, on the other hand, is rather well defined, giving the whole formation the appearance of being tilted to one side. The interior floor of Yerkes is nice and smooth. Wrinkle ridges approach the ring's ramparts from both north and south, but interestingly do not continue into the interior, which is graced by a minute hill; a nice string of hillocks is visible north of the formation's northern wall. Immediately west of Yerkes are the **Lavinium** and **Olivium Promontories**, bright protrusions that catch sunlight at about the same time as the west wall of Yerkes itself. Figure 4.1 is a highly accurate drawing of Yerkes and its environs.

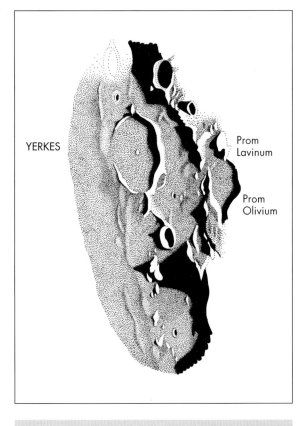

YERKES

Prom Lavinum

Prom Olivium

Figure 4.1. Yerkes and its environs. *Illustration by Nigel Longshaw.*

Peirce is the second-largest crater on the Crisium plain, only 19 km (12 miles) wide, with mountains 2100 m (7000 ft) high serving as its boundary. To its north is the much smaller crater **Graham**, which is connected to Peirce by a ridge. Near first quarter it can be difficult to distinguish these two features.

Firmicus, northeast of **Apollonius** (which is just beyond the southern boundary of mare Crisium), is another ringed plain, 56 km (35 miles) wide. It has a very dark floor which holds its contrast well even under high Sun angles. There is a hillock in the center, and a craterlet just inside the east wall. A valley can be seen just outside this wall, while on the opposite site of Firmicus lies a crater chain.

Cleomedes, named for an ancient Greek astronomer who may have been the first to discover atmospheric refraction, is a wonderful walled plain about 130 km (80 miles) in diameter, bordered by extremely broad mountains that rise more than 2500 m (8500 ft) above its floor. The very thick ring formed by these mountains is worthy of detailed study, for there are many features within the ring itself. There are numerous undulations and depressions in the crest of the eastern wall, while the west side is equally interesting, pocked with many craters and craterlets, many cocked at odd angles, and with very complex terracing. Look also for a deep, dark gorge that looks like a comma just south of east along the interior of the eastern wall. Just northwest of its center is an interesting group of peaks, south of which lies the crater **Cleomedes B**. Northeast of these peaks, a little less than halfway toward the northeast wall of Cleomedes, begins a Y-shaped rille that runs southeast. One branch of the Y runs to the base of the east wall of Cleomedes; the other, westernmost branch, curves southward and terminates just inside the southern wall. This wall, which merges into the northern boundary of Mare Crisium, is capped by a large crater.

Figure 4.2 shows three different views of Cleomedes, from observations made with a 75 mm (3-inch) refracting telescope. The drawings show the walled plain under different angles of illumination. The appearance of Cleomedes changes markedly even with small changes in the Sun's altitude above it. Figure 4.2 demonstrates why it is profitable to study lunar features over and over again at different angles of illumination or co-longitudes – you will always see something different from the last time

Burckhardt, northeast of Cleomedes, is a somewhat unusual-looking 56 km 35-mile wide ring bounded by very high mountain ridges to the east and west – those on the west side rise to altitudes exceeding 4000 m (13 000 ft). Keep in mind that on the Moon, which is a much smaller body than the

Earth, this is a very great height relative to "sea level." Wilkins and Moore described these curved mountain ridges as "possibly the remains of once-complete rings which existed prior to the formation of Burckhardt," but Goodacre says of them, "The walls are continuous … " How do you see them? Burckhardt also has the usual central hilly area, and other features in its interior that are visible in amateur-sized instruments, including a curving ridge inside the east wall and two crater pits to the southeast. **Burckhardt A** lies to the southwest, and east of the main crater is a chain of smaller craterlets.

Geminus. is a 90 km (55-mile) wide crater whose very broad walls are an excellent illustration of the terracing effect. It is truly a challenge to make an accurate drawing of all of the terraces and "shelves" that can be seen inside this crater's walls. The walls are quite lofty – rising to a height of some 3500 m (12 000 ft) on the west, and even higher, nearly 5000 m (about 16 000ft), on the east. In the center of the crater is a short curved mountain chain with at least three peaks. A broad gorge runs east-southeast from just outside the crater wall: this is called **Rima Burckhardt**, since it runs generally in the direction of that crater. The area all around Geminus is richly cratered, with many hilly regions. Look to the west of Geminus and you will see a key-shaped formation – this is **Geminus B**, whose north wall is completely broken open, leading to an elongated depression that forms the "teeth" of the key. **Geminus A** is a more ordinary-looking crater with an irregular border that lies just northeast of the "key" formation.

Bernouilli is a nice 40 km (25-mile) diameter crater with high walls, rising to nearly 3500 m (around 11 000 ft) on the west side, 4000 m (13 000 ft) on the east side. There are one or two hills visible on the crater's floor, and also a few hills just inside the west wall. A small craterlet appears towards the north margin of the crater. Beyond the eastern wall are a number of eroded crater rings, and more lie outside the southern wall.

Gauss, one of the largest of the Moon's walled plains at more than 160 km (100 miles) across, is an amazingly complex feature You could spend many nights exploring Gauss – and indeed you should! The walls of this crater have a very complex structured and are broken or eroded in many places. **Gauss A** sits atop the east wall, while the much larger crater **Gauss B** lies immediately to its southwest, just inside and intruding into the eastern wall of the principal crater. A deep rille runs from the northern edge of Gauss B toward the northeastern rim of the main crater, but ends in a hilly area before it can quite get there. It is only one of many rilles that can be glimpsed on the floor of Gauss;

Figure 4.2. Three views of Cleomedes, under slightly different angles of illumination. *Illustrations by Sally Beaumont.*

most of them appear to radiate from a point at the northern edge of the crater, the longest running all along the inside of the western wall. Another prominent rille runs also from the north rim southward, just east of the aforementioned rille, halting at a craterlet known as **Gauss E**, then starting up again and continuing southward all the way to the south rim. There is also a curved chain of mountains just west of center that runs north and slightly east; the mountain nearest Gauss's center appears to have a craterlet deforming its peak, and it is the brightest object on the floor of the main crater.

The region just south of the northern rim of Gauss is very complicated – a series of gorges and hills, a heavily scarred area indeed. A grouping of four small craterlets is visible just inside the north rim, which is pointy. To their east is a much larger

craterlet whose south rim has been obliterated, giving it the appearance of a horseshoe. **Gauss F** is a crater with a second crater inside it, just south of the central mountain ridge, while **Gauss G** is an odd, teardrop-shaped crater just inside the south wall. Its southwest wall is missing. **Gauss W** is a nice oval-shaped crater just inside the southeast rim. Look closely at the south walls of Gauss, and you should be able to make out several "passes" through the highlands here, some of which are quite broad. The area all around Gauss is equally rugged, containing innumerable gorges, rolling hills, and of course craters of all different depths, some shallow and ill-defined, others deep and in sharp relief.

Strüve is a small crater ring with a central peak located in a darkened area that allows it to be seen even at high Sun angles.

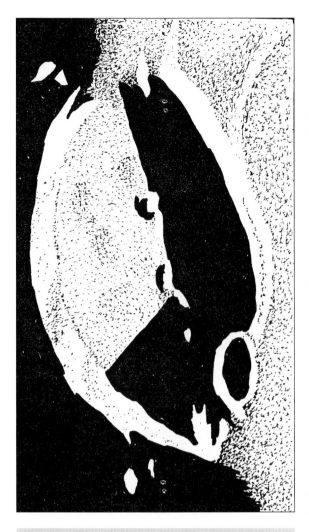

Figure 4.2. *continued*

Endymion is one of the larger formations in the Moon's northeast quadrant, measuring 125 km (78 miles) across; it lies fairly close to the Moon's northeast limb, just southwest of Mare Humboldtianum. Its walls, which rise more than 4500 m (15 000 ft) above the lava-flooded floor, are actually very wide, but are made to appear less so by the foreshortening that occurs near the Moon's limb. The floor of this crater is very dark and smooth as flows have obliterated almost all the floor features that used to exist here. The dark lava is streaked here and there with smeared whitish lines and smudges. The north edge of the crater contains three tiny craterlets all in a perfect line. The whole western wall is pocked with craters, some large, many overlapping one another. A nice chain of four large craters lie just outside the eastern wall, and to the northeast a mountain ridge extends between

Endymion and Mare Humboldtianum. Some gorges may be glimpsed along the southern wall.

Atlas, along with its smaller partner Hercules, forms one of the nicest crater pairs on the Moon's nearside. Atlas has a diameter of 89 km (55 miles), with walls that vary between 2750 and 3500 m (9000 and 11 000 ft). It has an extraordinarily complicated floor, crisscrossed by rilles and hills, and dotted with craterlets. At its center is a grouping of hills that give the impression of a ruined crater ring its eastern edge bordered by the deepest and most prominent rille inside the crater. Inside the south wall, this rille joins another that runs northwest, where it meets yet another running alongside the inner west wall. There are many variations in the elevation across Atlas's floor, which may account for the "glittering effect" noted by Wilkins and Moore under conditions of high illumination. **Atlas A** lies to the east; it has a prominent central peak. To the north is a large, eroded ancient ring, **Atlas E** (also called **O'Kell**); to the west is Hercules.

Hercules, at 72 km (45 miles) across though appearing somewhat smaller than that, is the twin of Atlas. Its heavily terraced walls rise 3500 m (11 000 ft) above its floor, but foreshortening can make the terracing difficult to discern. Inside the eastern walls are some landslips and a ravine that terminates in a crater. An oval crater intrudes upon the south rim, and from it emanates another ravine that zigzags along inside slope of the western wall. The floor is fairly smooth, but a large crater dominates the southern half of the interior. Outside the west walls is a large dark terrain known as **Lacus Mortis**, the Lake of Death.

Figure 4.3 is a pair of drawings of Atlas and Hercules at different stages of sunset, made with a 75 mm (3-inch) refractor. They show that amateur astronomers can make rewarding observations of the Moon with very modest-sized instruments. In the first drawing the floor of Hercules is half in shadow; in the second drawing it is completely in shadow. Atlas E shows up as a white ring in the first drawing, below and between Atlas on the left and Hercules on the right.

The second drawing is even more interesting. Now only the west edge of Atlas can be seen, but **Atlas E (O'Kell)** appears as a bright white ring with a very black interior and a whitish central peak catching the last rays of sunlight. A close examination of Lunar Orbiter photographs of Atlas E fails to disclose the existence of this mountain peak, but it is doubtless a real feature, and I have seen it many times. This clearly refutes the mistaken idea, held by many amateur astronomers, that there is nothing to be discovered on the Moon which has not already

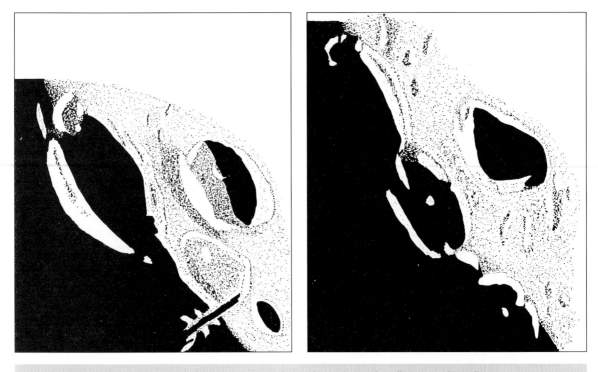

Figure 4.3. Two views of the twin craters Atlas (right) and Hercules. *Illustrations by Sally Beaumont.*

been adequately photographed by the many spacecraft sent there from the 1960s.

Franklin, named for Benjamin Franklin, is 55 km (34 miles) across with walls 2500 m (8000 ft) or so high, with prominent terracing. There is a grouping of central hills, with a high ridge running south from them; just west of these hills is a blotchy bright terrain. You can also make out a group of craterlets along the inner eastern wall. Just outside the southern rim is **Franklin S**, a partly ruined walled plain, its northern half gone; immediately south of this ruined crater is a chain of large craterlets, some very deep and dark.

Democritus is a deep crater ring 40 km (25 miles) wide with 1500 m, (5000 ft) high walls and a prominent central mountain peak. **Democritus A** is a craterlet close to the south wall; nearby is **Democritus B**. Together these two craterlets mark out the site of an ancient ring that is now highly eroded.

Mare Humboldtianum, Humboldt's Sea, lies at the extreme northeast limb of the Moon. It is a pear-shaped plain, foreshortened to a greatly elongated oval by its location. This is a "libration object" – a lunar feature that the observer will better see when the Moon is turned a little to the east to show us a small part of the farside that is ordinarily hidden from view. Even at lunations with little libration effect, you will be able to make out a lengthy

chain of mountains that forms the western boundary of this mare; these mountains are riddled here and there with craters, and there is a nice lava-filled ring towards the northern end, just inside (east) of this range. A much larger lava-filled ring lies to the south; its southern rim broken by two passes leading out onto Humboldtianum's plain, which is in general greatly smoothed over by its own lava flows. Three lunar domes can also be spotted along the western edge of the formation. The southeast wall has been described by Wilkins and Moore as consisting of "concentric ridges"; the description is apt.

Figure 4.4 shows Mare Humboldtianum. Notice the many different subtle shadings, which indicate different albedos for the various features on the floor of the lava plain.

The **Taurus Mountains**, located between Mare Crisium and Lacus Somniorum, climb to 1000–3000 m (3000–10 000 ft). This area of the Moon appears very dark, even at full moon, and trying to sketch all of the rounded peaks that can be detected under high magnification can be quite a challenge. Using low powers to observe this region, you will probably discover the presence of several rilles near the Taurus Mountains that have a bowed-out appearance; higher powers clarify these intricate carvings in the lunar surface, which probably are

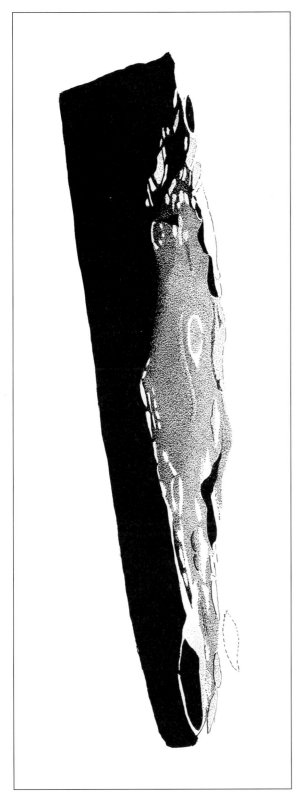

Figure 4.4. The highly foreshortened Mare Humboldtianum. *Illustration by Colin Ebdon.*

shock waves from the tectonic stresses in Mare Serenitatis that built up as ancient lava flows filled it in. The last manned Moon mission, Apollo 17, explored the Taurus Mountains.

Posidonius is a very large (100 km, 62 miles, wide) and detailed walled plain, a favorite object in the northeast quadrant for lunar observers. This crater was almost certainly dug out by an impact that occurred after the much larger one responsible for the Mare Serenitatis basin, but before the extensive lava floods that later filled Serenitatis. Some of these lava flows have obviously affected Posidonius, intruding into what would otherwise be a more perfect structure. The walls are only of average height (no more than 2000 m, 6000 ft), but are interesting for their uneven thickness – the walls on the east are very wide, but very narrow on the west side. One of the most striking features of Posidonius is the presence of a second walled plain lying inside the main outer wall, giving the formation the appearance of concentric rings. The inner ring is not quite perfect – the southwest and parts of the north rims are broken. This inner ring is almost perfectly bisected by a beautiful rille, running northwest–southeast, tangential to the east rim of Posidonius A (see below). This rille was observed in 1929 by Russell W. Porter, who used the Mt Wilson 60-inch (1.52 m) reflector; but do not despair – you don't need such a magnificent instrument to see this rille well, as it is striking in my 250 mm(10-inch) telescope!

The floor of Posidonius, which has been described as "glittering," is very bright. At the center of the whole formation is a bright peak, the first and largest in a semicircle of mountains (perhaps the remnants of yet another "crater within a crater") that lie due east of **Posidonius A**, a perfectly round crater also very near the center. Smaller craters can be found in the southern interior, while the north regions of the floor are hilly. A low ring and several craterlets may also be found here. **Posidonius J** is a fairly deep bowl carved out of the north rim, with the smaller **Posidonius B** and **D** lying southeast of it, intruding onto the northeast rampart. A fine rille may be found on the southern part of the interior floor, running southwest–northeast and making an inverted T with the longer rille mentioned above, and there is a very long rille alongside the inner west wall, also approximately northwest–southeast.

A wide gorge cuts across the south floor of Posidonius in a roughly southwest–northeast direction, southeast of and parallel to the fine rille described just above that forms the cross in the T. There are other gorges cutting their way through the north and east walls. Outside the southeast wall and touching Posidonius is the walled plain **Chacornac**, while beyond the southwest wall is an expansive, ancient ring-plain, from which emanates a low but rugged ridge of hills that curves southwest across Mare Serenitatis. This ridge attains heights of 250 m (800 ft) in places and is not a difficult object in amateur instruments. Figure 4.5 depicts Posidonius and some of the objects discussed above.

Littrow is a deep, 35 km(22-mile) wide crater whose western wall borders Mare Serenitatis. In the nineteenth century the amateur lunar observer Thomas Gwynn Elger described it as looking like the letter D; and Julius Schmidt, who did much of his work at Athens, Greece, saw Littrow as shaped like a pear and elliptical. In his classic 1931 guide *The Moon, With a Description of its Surface Formations*, Walter Goodacre pointed out that Littrow really is circular, appearing elliptical because of its proximity to the Moon's eastern limb. Its walls are very broad, especially on the west side; the southern boundary is broken up by a pass carved through it. Outside the eastern wall is a smaller, partly ruined ring plain, its north wall absent. Still further east lies a second, larger ruined ring. Look also for a chain of craters

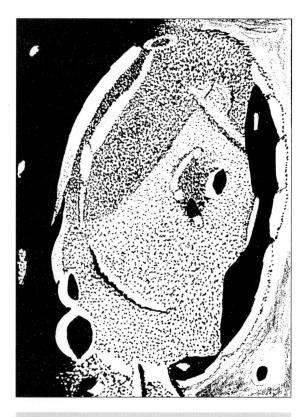

Figure 4.5. The concentric structure of Posidonius. *Illustration by Sally Beaumont.*

than runs down the slope of the outside north wall of Littrow. There is not too much detail on the floor of this crater, but some hills and mounds are detectable, as well as some fine albedo features.

Maraldi is a walled plain whose borders are rather uneven, greatly eroded at their southern part where the formation opens onto the surrounding lunar terrain. There are many ruined or partly ruined crater rings similar to Maraldi in its vicinity, testaments to the extensive lava floods that must have occurred there after the original craters were formed. The inside of Maraldi's northern wall contains a prominent craterlet; nearby, just outside the north wall, is **Maraldi M**, in the words of Walter Goodacre a "low bubble-like hill", which has been identified by some observers as a lunar dome. The floor is very dark indeed, and, according to Goodacre, "greatly depressed below the plain." **Maraldi S** and **P** are faint, lava-submerged rings that lie on either side of the gap in the south wall of the main crater. The still larger ruined ring **Maraldi R** is farther south of these features.

Mount Argaeus is an isolated mass of mountains located on Mare Serenitatis, found between the craters Littrow and Dawes. A rille runs from Littrow past the base of these mountains, which rise to 2500 m (8000 ft). At sunrise on the Moon, Mount Argaeus casts a remarkable pointed shadow. A ravine divides these mountains into two halves.

Macrobius is a walled plain 68 km (42 miles) wide, easily visible even at full moon. Its walls rise from 3000 to 4000 m (10 000 to 13 000 ft) above its floor, are terraced, and appear very bright. The north side of the walls has straight sections and contains a gap just east of due north. The terraces are unusually wide, and appear to repeat in concentric arcs right down to the central regions of the crater's floor. **Macrobius A** is a large craterlet sitting atop the crest of the southwest wall, and it is as bright as the wall itself. Just east of center is a grouping of three hills, just a few of many hills and domes on this crater's floor. Look also for a ravine running along the outer northeast wall to the crater **Tisserand**. Outside the northwest rim is a low, mostly ruined ring.

Proclus is a small (only 29 km, 18 miles) but very deep and bright crater characterized by its extensive system of lunar rays. Like all lunar rays, they are brightest at full moon. The walls of Proclus, which are broken by alternating bright and dusky bands, rise to the surprising height of 2500 m (8000 ft) above the floor of this diminutive crater, and appear very white, even at fairly high Sun angles. Proclus is fairly round, though its location on the lunar surface foreshortens it into an ellipse; its southwest walls

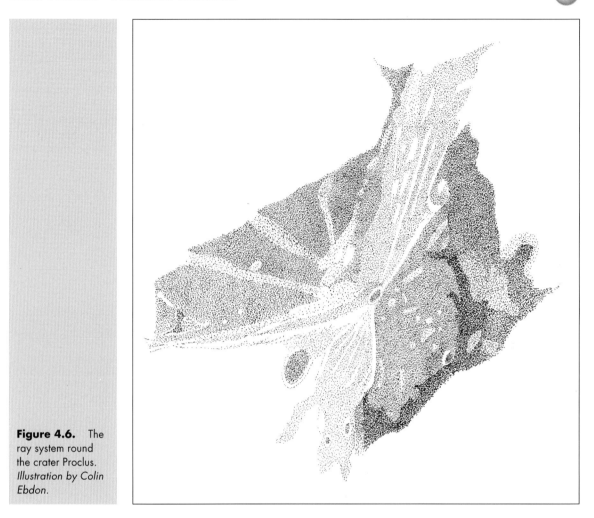

Figure 4.6. The ray system round the crater Proclus. *Illustration by Colin Ebdon.*

are squared off, and indented by a craterlet. There are a few hills on the floor, but details here are difficult to see owing to the overall brilliancy of this object, which washes out most of the subtler features inside it.

Figure 4.6 is a drawing of the ray system of Proclus. The brightest ray, which runs northwest, appears to follow a shallow ridge at the edge of Palus Somnii. Finer rays radiate to the north, while a long, curved "double" ray extends far to the northeast. One of these has a "feathered" appearance. These double rays crisscross one another at one point, and at another point they appear to be joined by a subtle filament. The longest of the rays, which can be traced south all the way to the "shoreline" of Mare Fecunditatis, is just one of many radiating southward. A series of very broad, paler rays fans out to the east, across Mare Crisium. Interestingly, no rays extend westward across Palus Somnii. Lunar scientists have questioned whether the ray system of Proclus actually originates from Proclus itself or

from other, closely-associated craterlets; it has even been proposed that the rays may be classified into three separate families, radiating from three different points near the main crater.

Palus Somnii, the Marsh of Sleep, is located southwest of Mare Crisium and is easy to see under any Sun angle as it is dark with well-defined edges. Its surface is not as smooth as many of the other lunar regions designated as "marshes", which are just lava-covered areas that are not as large as the maria, or "seas." These marshes often lie near the perimeters of the seas, so perhaps their nomenclature is not so inappropriate after all. The western side of Palus Somnii borders Mare Tranquillitatis. The surface of Palus Somnii is strewn with hills, craters, and ruined ring formations such as **Lyell** and **Franz**. Lunar observers have noted colors in the region, but these are probably just the optical effects of changing conditions of illumination.

Plinius is an almost perfectly round crater, 48 km (30 miles) wide, encircled by broad, finely terraced walls that spill their material onto the surrounding, darker terrain well outside the walls of the crater. Besides the very complex and beautiful structure of these walls, the most striking feature of Plinius is its central mountain peak, which is capped by a crater, much like volcanic calderas on Earth. This central "mountain" may simply be the rim of an ordinary crater, placed perfectly at the center of Plinius. An isolated mountain lies just southwest of this central formation, while another isolated mountain is located just inside the northwest wall. A long rille runs along the lunar surface a little north of the north wall, toward the crater Dawes. Figure 4.7 is a beautiful drawing of Plinius and the region to its north.

Menelaus is a 32 km (20-mile) diameter crater which, like Plinius, has well-terraced walls that soar 2500 m (8000 ft) above its floor. The northern walls

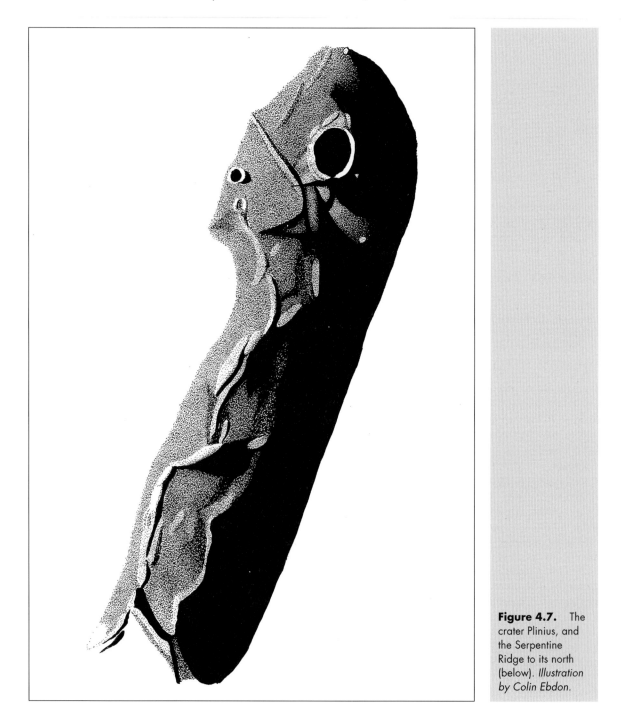

Figure 4.7. The crater Plinius, and the Serpentine Ridge to its north (below). *Illustration by Colin Ebdon.*

are broken by lava flows from Mare Serenitatis. Menelaus is very bright at full moon; from it radiates a streak of bright material which crosses Mare Serenitatis. This streak includes a rille which is about 30 km (20 miles) long. West of Menelaus lie the Haemus Mountains. **Menelaus A** lies northeast of the principal crater. The floor of Menelaus has a mountain that's a little off-center, with a ridge extending from it to the inner southwest wall.

Mare Serenitatis, the Sea of Serenity, is 690 km (430 miles) wide and a perfect example of the Moon's great lava plains. Its circular shape betrays its genesis in a giant asteroid impact almost 4 billion years ago; the lava flows came later. The Taurus Mountains border this mare on its eastern side, while the Caucasus Mountains form the northwest border, with the Haemus Mountains to the south and west and the Posidonius uplands to the northeast. The eastern shores of Serenitatis are dominated by a wrinkle ridge that runs north-south along almost its entire length, rising to heights of 250 m (800 ft) in places. This feature, shown in Fig. 4.7, has been dubbed the **Serpentine Ridge** because of its snake-like undulations. The best times to observe this wrinkle ridge, the largest on the Moon, is either at the five-day old Moon (sunrise) or at the eighteen-day old Moon (sunset). Like the other maria, Serenitatis has a fairly featureless floor (who knows what features lie buried underneath all that lava?), but the margins contain some interesting objects.

Linné is only 1.5 km (1 mile) across, so you might well wonder why we should bother to mention this tiny crater. Located on the western shore of Mare Serenitatis, diminutive Linné is characterized by a mysterious white "collar" that encircles it. This whitish material makes Linné easily visible against the dark mare basalt material. It's interest lies in its very controversial history: in the past, many reports of changes in its appearance convinced observers that they were looking at a lunar volcano in action. The most amazing thing about Linné is the fact that at low angles of solar illumination it is nearly invisible, except with large telescopes under very high magnification, and then only if you know exactly where to look. But as the Sun climbs higher above it, the while collar around Linné brightens and expands the apparent diameter of the crater to four times its normal size, making it prominent. Eventually the inner crater itself is swallowed up by the white ring around it. Linné is one of the youngest lunar craters known – "only" a few tens of millions of years old, in contrast to the billions of years in which the ages of most craters are reckoned.

If you need convincing that you must observe the Moon constantly, from one lunation to the next, in order to fully appreciate the different aspects of its surface features, look no further than Linné! From one lunation to the next, the aspect of the Sun's rays on the lunar surface will change slightly, showing up different aspects of the same feature – usually these differences are very subtle, but with Linné the changes can be more significant. On occasion, the changes can be downright startling, especially to the neophyte observer who is used to the comparatively unchanging face of the Moon. It is no small wonder, , that many lunar observers of the past hypothesized that these changes were being caused by volcanic eruptions that spewed fresh white ash over the surface.

Modern spacecraft have investigated Linné, failing to find any evidence of volcanic activity. Instead, the changing appearance is explained by variations in the angle at which sunlight is reflected from the whitish blanket of ejecta that makes up the collar around the crater. But Linné still offers the lunar geologist plenty of mystery. Why, for example, does this crater have a nice, perfectly round collar of bright ejecta material, while other, similar craters have ray systems instead? Our current understanding of the mechanics of impacts (factors such as the composition and speed of the impacting body, and its angle of impact) do not adequately explain the different outcomes, so there might be a better explanation.

The **Caucasus Mountains**, which divide Mares Imbrium and Serenitatis, contain a wide variety of steep, lofty peaks – the highest of which reach 6000 m (almost 20 000 feet – and low hills. Ejected material from Mare Imbrium has scraped across the Caucasus, which run in a roughly north–south direction. Selenographers have identified five distinct sections of the Caucasus. The most southerly section is mostly made up of isolated mountain peaks. The section immediately to its north is more plateau-like, separated from the first section by a series of valleys and bordered at its north end by a single wide valley that runs across its whole width to the crater Theaetetus. The middle section is the largest of the five and contains the bright, 32 km (20-mile) wide ring Calippus; the widest section of these mountains occurs at a point south of Calippus, where they extend east–west for just over 100 km (65 miles). The fourth section extends to Eudoxus and consists of further isolated peaks and masses. The fifth and northernmost section is formed by a semicircle of mountains that roughly connects Eudoxus to Aristoteles.

Eudoxus is a highly symmetrical, 72 km (45-mile) wide walled plain with walls that rise as high as 3500 m (11 000 ft) above its floor, and a high degree of terracing visible along its inner ramparts. The west wall is cut by a gorge which separates the rest of the wall from its crest. Some observers see the west wall as double for this reason. The walls to the north and southeast are also disrupted by what appear to be valleys radiating outward from the crests of the walls. **Eudoxus B** is a bright crater just beyond the north wall, or perhaps on its slope. **Eudoxus A** and **G** are situated to the east of the principal crater. The former is quite sizable for a secondary crater, measuring 14 km (9 miles) across. Around 65 km (40 miles) west of Eudoxus is a curved mountain ridge; nearby are **Eudoxus C** and **D**. North of this ridge, the astute observer will be able to make out one or more rilles. The floor of Eudoxus is unremarkable, save for a few modest hills in the southern regions.

Aristoteles, located south of the Mare Frigoris, is one of the Moon's larger walled plains, measuring 97 km (60 miles) across, with greatly disturbed walls cut by massive canyons and ravines. Like so many walled plains, Aristoteles shows fine examples of the terracing phenomenon inside its walls, and accurately drawing all of these landslips can be difficult, especially for the beginner. In some parts the walls rise to altitudes of 3500 m (11 000 ft) or so; one can only imagine the vistas from these peaks! Many hills and domes are scattered across the floor of this crater.

The **Alpine Valley**, visible from first quarter as a darkened gap through the Alps Mountains, is a lava-filled graben – a unique feature of the northeast quadrant. Remember from our crash course on lunar geology that a graben is formed when a width of land sinks between two fractures or faults that run parallel, or approximately parallel, to each other. In the case of the Alpine Valley, the two faults are just two of many such fractures that appear to radiate outward from Mare Imbrium, a giant impact basin discussed in the next chapter. Like other maria, Imbrium was created by a very powerful impact that took place about 3.85 billion years ago. The force of this impact caused numerous fractures in the Moon's crust, many of which are visible in the Alps Mountains and all along the perimeter of Imbrium itself.

At first quarter, not much of the floor or walls of the Alpine Valley can be seen, but after the Moon is seven days old a surprising amount of detail can be made out within the formation. The valley, which is about 120 km (75 miles) long and 6 to 10 km (4 to 6 miles) wide, appears to begin inside the highest regions of the Alps Mountains themselves, in a nearly round solidified lava pool that Wilkins and

Moore aptly describe as an "amphitheatre." From there it cuts through the lower Alps and their foothills, taking on the shape of a narrow cigar. Some of these foothills are actually volcanic cones.

With but a couple of exceptions, the floor of the Valley is essentially smooth and featureless, because it is lava-filled, like the maria. Goodacre described it as "uniformly dark and similar to that of the Mare Frigoris." Two faults cut across the valley floor at roughly right angles to its length; if you have a very sizable and well-figured, precisely collimated telescope, you might be able to make out the very fine sinuous rille that meanders down the central floor. William H. Pickering of Harvard discovered this rille. I have seen it well with the 40-inch (1 m) refractor at Yerkes Observatory, but have glimpsed it in my 250 mm (10-inch) f/13.5 telescope at high power (×686). The rille appears to originate from a vent crater, which is typical of sinuous rilles in general, suggesting their volcanic nature. It is probably nothing more than a collapsed lava tube. The walls are more interesting, and they stand in contrast to each other. The west wall is rough and irregular, but the east wall almost uniformly linear in appearance.

The **Alps Mountains**. are probably the second most significant mountain range on the Moon's surface, next to the Apennines. These mountains curve along the northeast rim of Mare Imbrium for a distance of about 290 km (180 miles) between Plato and Cassini. The range contains many peaks that reflect sunlight brightly, so that it can often be difficult to make out details in the mountains themselves. Most of these peaks rise to heights from 2000 to 2500 m (6000 to 8000 ft), but there are a few that are much higher – 3500 m (nearly 12 000 ft) in the case of Mont Blanc, one of the Moon's greatest mountains. East of Plato is a 3500 m (12000 ft) high peak. The Alps Mountains are traversed by the famous Alpine Valley.

Palus Nebularum, the Marsh of Mists, is an extension of Mare Imbrium to the east of Aristillus and Autolycus, and has a somewhat lighter tint that the rest of the mare. A fairly featureless region, covered with solidified lava, Palus Nebularum does show several hills and ridges, and a few small craterlets.

Palus Putredinis, the Marsh of Decay, southeast of Archimedes, is another bay of Mare Imbrium. Like Palus Nebularum, this "marsh" contains a few hills and ridges, but also shows a series of rilles, including the **Hadley Rille**, a volcanic lava tube visible at last quarter.

Cassini is a 58 km (36-mile) wide crater just south of the Alps Mountains. Lava from the Imbrium plain has clearly intruded into Cassini, where it has obliterated many features that would doubtless be visible within the crater; however, unlike many other ruined craters, Cassini's ejecta just outside its rim is still very much apparent – at least some of the ejecta was piled up high enough to avoid being completely drowned in lava flows. However, not too much of Cassini's rim rises above the flows, so that you must observe this object when it is near the terminator, as its low contrast at any other time washes it out – Cassini virtually pulls off a disappearing act at full moon. While the east rim rises to 1200 m (4000 ft), the west and north rims are much lower – only 450 and 250 m (1500 and 800 ft), respectively. The well-formed rim of Cassini and the ejecta blanket just outside give the crater a double-rimmed effect.

Two sizable craters are contained within Cassini: **Cassini A**, a 13 km (8-mile) wide feature dubbed "The Washbowl" by Wilkins and Moore, and the more diminutive **Cassini B**. A still smaller craterlet lies southeast of Cassini A. The keen-eyed may be able to spot a couple of low hills between Cassini A and Cassini B. **Cassini M** is a crater that lies atop the northwest rim. In 1952, using a 320 mm ($12\frac{1}{2}$-inch) telescope, Patrick Moore discovered a very shallow whitish ring situated just south of Cassini and almost touching its south rim. Photographs barely show this ghostly feature. Can you see it? Figure 4.8 is a drawing of Cassini, its floor in deep shadow.

Theaetetus is a 26 km (16-mile) wide crater often seen in high relief, since its walls soar 2000 m (7000 ft) above its bottom. One reason for this great height differential is the fact that the floor of Theaetetus is 1500 m (5000 ft) lower than the surrounding terrain, so the bounding mountains are not as high as one might think; rather, Theaetetus is a very deep hole with a modestly high rim. Lunar transient phenomena (LTPs) have been observed in and around Theaetetus; whether these observations were of real events like small-scale outgassings or whether they were optical effects remains hotly debated to this day.

Aristillus forms a pair with the walled plain Autolycus on the eastern shore of Mare Imbrium. Aristillus is 56 km (35 miles) wide with thick, high walls: 2750 m (9000 ft) on the east side, 3500 m (11 000 ft) on the west side. Much of the inner walls are richly-terraced, but the outside walls are even more interesting, displaying many canyons of considerable depth. A ghost crater appears to just touch the north rim. The south rim is undulating, sometimes giving the appearance of a lasagna noodle! The floor is decorated with a magnificent central mountain mass, consisting of at least three peaks and several lower plateaus. Like Theaetetus, the floor is well below the surrounding lunar terrain – 1000 m (3000 ft) . Look for the complex series of ridges that radiates outwards from Aristillus across Mare Imbrium in all directions, but particularly north and south, like spokes on a wheel. Many of these ridges show a definite curve to them. They are most probably ejecta from the impact that created Aristillus. A faint ray system emanates from the crater.

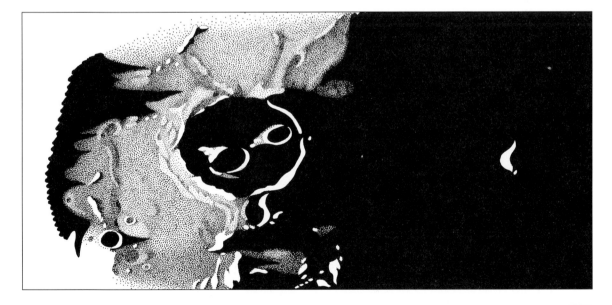

Figure 4.8. Sunrise over Cassini, with "first light" on Mt Piton. *Illustration by Nigel Longshaw.*

Autolycus is smaller than Aristillus, and very different from its neighbor: 39 km (24 miles) across with walls 2750 m (9000 ft) high, Autolycus is far less complicated, having thinner walls with little terracing and no big mass of mountains at its center, though a very interesting group of rounded hills lies south of the crater. The rim comes to a point in the south, and on the west side it is dimpled by a deep craterlet known formally as **Autolycus A**. Two very shallow craters appear on the otherwise unremarkable floor, the northernmost of the pair very elongated.

The **Apennine Mountains** are easily the most impressive of the Moon's mountain ranges, forming the 675 km (420-mile) long southeast border of Mare Imbrium. Its loftiest peak is **Mt Huygens**, which towers 5600 m (18 500 ft) above the mare plain, but there are many other extremely high mountains in the range, like **Mt Bradley** (5000 m, 16 000 ft), **Mt Hadley** (4500 m, 15 000 ft) and **Mt Wolff** (3500 m, 12 000 ft). A number of the peaks in this range are capped by crater pits – Huygens has a craterlet very near its summit which is 1.5 km (1 mile) across. On the "seaward" side, the Apennines rise above Imbrium in great, steep cliffs. While these altitudes may not seem terribly high by terrestrial standards, we must keep in mind that the Moon is a much smaller planet than Earth, so that the relative heights of the Apennines are much greater than even the highest mountains on Earth, such as the Himalayas. There are many valleys and gorges that run through the Apennines – look, for example, for a prominent valley near the crater **Conon**. You might notice that most of these valleys run southeast away from Imbrium, a uniformity of orientation that Goodacre called "curious." But their direction is not curious at all, for they are likely shock fractures from the Imbrium asteroidal impact, similar to the fractures that formed the Alpine Valley.

The region of **Hyginus** is one of the most rewarding lunar formations to observe with a small telescope. Hyginus itself is just a crater pit no more than 6.5 km (4 miles) in diameter, and would otherwise not deserve mention in any observing guide. Its north wall is broken by a craterlet that is a third as large as Hyginus itself. But Hyginus is of course famous for the dramatic rille that is roughly centered on the crater. The **Hyginus Rille** is shaped like a flattened letter V, with Hyginus at the bottom point of the V. The rille is easily seen in a 60 mm ($2\frac{1}{2}$-inch) telescope under steady skies. Numerous craterlets pit the floor of the rille all along its length, widening it at many points. The eastern section of the rille is the longer, starting from a craterlet slightly north of Agrippa. The western section terminates in a valley leading towards Mare Vaporum. The valley is just east of a mass of mountains. Look also for a very delicate rille that runs from the southwest rim of Hyginus crater to the rille system of Triesnecker nearby. Other rilles emanate from the southern rim of Hyginus through a dark spot about 13 km (8 miles) across. This spot may actually be a shallow crater itself. What do you see here? To the north of Hyginus and connected to it by a wide, shallow valley is the feature Schneckenberg, which Wilkins and Moore call "a curious spiral mountain."

Ariadaeus is a bright but unremarkable little crater, only 14 km in diameter, but the **Ariadaeus Rille** running northwest of it onto the Mare Vaporum is one of the most beautiful on the Moon's surface. Johann Schröter discovered this graben in 1792, but it can easily be seen in almost any amateur's telescope. The great rille arcs for over 250 km (150 miles). It is forked at its eastern end, next to Ariadaeus; the north fork is broken by a mountain. The rille is "fractured" at roughly its midpoint, south of Julius Caesar. This is what geologists call a strike-slip fault. The west end is also forked, where it joins another great cleft, the Hyginus Rille.

The terrain across which the Ariadaeus Rille runs is far from smooth, but the numerous hills and crater rings don't seem to bother the rille one bit – it runs roughshod over all of these smaller features. Since Ariadaeus Rille is located near the center of the Moon's nearside, it is best seen from a couple days before first quarter to a couple days afterwards. It is nearly invisible at full moon. Ariadaeus Rille is best known for the fact that you can see features on its floor; for most rilles, it is impossible to do this with amateur instruments. These features include several craterlets and depressions. Figure 4.9 is a drawing of the Ariadaeus Rille. Note the large ruined ring south of the rille at center, and the several hills and bluffs casting long shadows at sunrise.

The **Triesnecker Rilles** form one of the most intricate systems of rilles on the Moon, and the patient amateur observer should be able to see many different branches, which look rather like fractures in a cracked eggshell. The crater **Triesnecker** is itself a beautiful and intricate medium-sized walled plain 23 km (14 miles) across, rounded on its eastern half, polygonal on its western half, with a complicated group of central peaks and irregularly terraced inner walls. The rille system runs north–south over the flat lunar surface outside Triesnecker's east wall. The most prominent of the rilles come to a V that points toward their namesake crater. The rilles branch northward all the way to

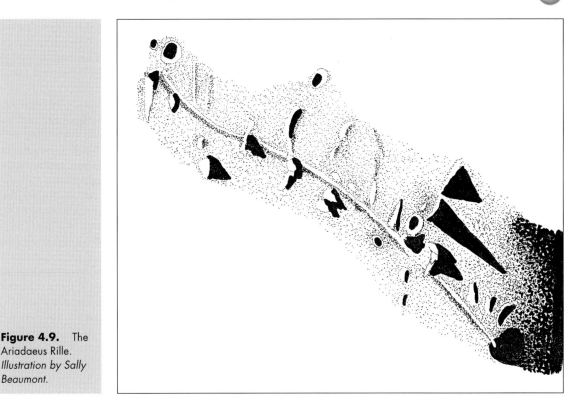

Figure 4.9. The Ariadaeus Rille. *Illustration by Sally Beaumont.*

Hyginus and southward to Rhaeticus. It is interesting that instead of rilles, the lunar terrain west of Triesnecker is deformed by many wrinkle ridges, also running roughly north–south.

Manilius is a wonderfully detailed walled plain 40 km (25 miles) wide with southern walls that are somewhat squared off. The crater is very bright under high Sun angles, and even has faint rays, though these can be challenging objects to see. A deep gorge cuts across the floor from the inside of the southern wall. Manilius has a very asymmetrical blanket of bright ejecta material, with almost all of the debris located beyond the eastern wall. The central floor has the usual mountainous terrain, with several craterlets at its base.

Julius Caesar. is a largely ruined ring plain, 90 km (55 miles) wide, located between Mare Vaporum and Mare Tranquillitatis, whose southeast wall has been completely knocked down. Most of the interior has, not surprisingly, been flooded with lava, and so you should expect to see only a few odd craterlets there. There is a plateau inside the northeast rampart, and the eastern half of the crater is generally rougher and more uneven than the western half. The northwest quarter of the floor is noticeably darker than the rest of the floor. Walter Haas, the founder of the Association of Lunar and Planetary Observers (ALPO), has noted this area as

having a "bluish" appearance during the lunar morning.

Boscovich is located in an elongated valley related to the Imbrium impact event, and in many respects is just a smaller version of the crater Julius Caesar. Its walls are also eroded, especially on the west side, and very dark lava has completely overwhelmed its interior. In fact, the lava flows have made the floor of Boscovich one of the darkest features on the whole Moon. Two very fine rilles cross the floor of this crater; the easiest to see runs across the entire floor in a north–south direction. Wilkins and Moore found variations in the coloring of the floor. There is a very wide opening in the southeast wall, which leads to a long ridge, one of several in the area, all running parallel to one another.

Dionysius is a 19 km (12-mile) wide crater located at the extreme southwest shoreline of Mare Tranquillitatis. Although the interior contains a few subtle features, including some landslips on the inner walls, the area outside the crater is more interesting. To the east lie a series of rilles, all oriented northwest–southeast, and named after the nearby crater **Ritter**. In contrast, the terrain to the west is very smooth and unscarred. Dionysius frequently appears to be encompassed by a bright ring of ejecta material.

Silberschlag is a bright crater 13 km (8 miles) wide just south of the Ariadaeus Rille. A mountain range runs along the west side of this crater, continuing north and south of it for some distance. It has been suggested that at least the southern half of this range is actually the wall of an ancient crater ring.

Agrippa is a walled plain, about 45 km (28 miles) across, characterized by rather steep terraced walls that usually appear quite bright. The walls rise to more than 2000 m (7000 ft) above Agrippa's floor. The wall on the west side is fairly straight, whereas the crater's east wall follows a regular curve. The west wall also appears to be broken into two sections. The outer structures of the walls are just as interesting as the insides, with evidence of old landslides apparent. The north wall contains a craterlet designated **Agrippa H** and a protrusion which Wilkins and Moore described as a "rocky spur." How do you see this feature? Beyond this protrusion are a pair of craters – from the westernmost of the pair you can trace the famous Hyginus Rille. The south wall is equally mysterious, crowned by a craterlet or possibly a "pass" through the wall. Goodacre, however, called this feature a "peak"! The floor of Agrippa is most interesting, and you could spend many nights just studying all there is to see here. The main structure is a bright central mountain peak, which might actually be "double"; it is, to say the least, a very complex structure. **Agrippa F** is a large and shallow walled plain to the east of the main crater, contained within the even larger and equally eroded ringed-plain known as **Tempel**, which is itself as large as Agrippa.

Godin, immediately south of Agrippa, is an egg-shaped walled plain, 40 km (25 miles) in diameter, featuring a very well-defined rim whose inner slopes are heavily terraced. It is interesting to note the different terraces that are visible during slightly different co-longitudes. The terraced walls are not steep, as you might expect, but slope gradually towards a prominent mountain peak in the center of the crater. Look for a littler peak poised between the central mountain and the northern rim. One of the most challenging objects for the lunar observer in Godin is a ruined ring plain that actually lies inside the main crater, just to the west of the central peak. This feature was discovered by Patrick Moore in April, 1953 while he was using the great 0.84 m (33-inch) refractor at the Paris Observatory's Meudon site. It is a very subtle feature, and will push your observing skills to their limits. The crest of the south rim is pocked by a craterlet. Beyond the northwest outer slope is a well-formed crater known as **Godin A**, surrounded by bright ejecta and with two smaller craters that abut it. A little bigger than Godin A is another crater just outside the south rim, known as **Godin B**. To the west of Godin lies a large, barely discernible ruined walled plain, just one of many such structures in this neighborhood.

Rhaeticus is a large walled plain with irregular, broken walls lying on the Moon's equator, south of Triesnecker and its rille system. The walls of Rhaeticus are of average height for a formation of this type – about 1500 m (5000 ft), but are greatly disturbed in many places. A crater ring intrudes into the southeast rim, and two more (one inside the other) on the north rim, inside which are two small craterlets. A deep ravine cuts into the south rim. There are gaps in the east and west walls, too. The rim is dotted in several places by craterlets. On the outside of the rim, numerous ridges of ejecta slope broadly to the lunar floor below. A lengthy rille runs southwest from Rhaeticus to the crater **Réaumur**. This rille is difficult to make out near Rhaeticus because it has to run through a rugged mass of mountains to that crater's southwest, but it is more visible after it emerges from these highlands. Even more interesting is the rille running north–south just west of the west wall of Rhaeticus, which is almost a continuation of the extensive Triesnecker Rilles. A crater chain runs northeast and appears to continue into the interior of Rhaeticus, which is accented with many hills but has no major central peak.

Chapter 5

Lunar Features – Southeast Quadrant

Hipparchus is a very large walled plain 160 km (100 miles) wide, with somewhat irregular, highly eroded borders that some observers say have a hexagonal shape. Hipparchus has been described as a miniature version of Mare Crisium. Because the mountains that ring this formation are not particularly high, Hipparchus's outline is recognized only when near the Moon's terminator, and is badly washed out by the Sun's light at all other times. Although the bordering mountains may rise as much as 1200 m (4000 ft) above the crater's floor, they are almost level with the ground that surrounds the exterior of the formation, so there is little contrast except under very low Sun angles. This is truly an ancient crater, left over from the Imbrium impact. Its southwest side is almost obliterated. Three distinct valleys break through the southeastern rim, further interrupting the outline of this crater. One of these valleys emanates from **Hipparchus G**, a fine bowl crater on the crest of the east wall. Even some of the larger crater rings inside Hipparchus, including a ghost crater very near its center, have broken walls, open on their north sides to the lava-smoothed floor. The north wall is fairly well defined by rows of mountains that look higher than they are because of their very jagged appearance. This is an optical illusion: they are not jagged at all, the "jags" you see are actually horizontal gouges from the Imbrium impact (see below), not vertical relief.

The floor itself, besides the ghost craters just mentioned, is the site of a younger, well-preserved 29 km (18-mile) wide crater called **Horrocks**. Its 2500 m (8000 ft) deep walls are distorted into an odd, irregular shape, but its rims are sharp, the inner slopes are terraced, and it contains a fine group of central peaks, all evidence of its comparative youthfulness. The central and western parts of the floor are flooded with lava, but a few small hills poke through. On the south floor look for a long ridge that runs southeast to the crater Halley, which itself has a very smooth floor. South of Halley runs a remarkable deep, wide gorge which ends at the east wall of Albategnius. East of Halley is a curious group of three large aligned craters which point towards the northeast – the first is a little larger than the second, the second a little larger than the third. Beyond the east wall of Hipparchus lie several ruined crater rings and ghost craters, some quite sizable.

Albategnius, best observed at first quarter, is a beautiful and complicated walled plain about 130 km (80 miles) across. Like Hipparchus, Albategnius is an old impact site showing signs of much erosion, but the process hasn't advanced nearly as far, so the outline of this crater is much easier to discern than that of its neighbor to the north. The walls of Albategnius are studded with huge peaks that rise some 3000–4250 m (10 000– 14 000 ft) above the floor of the plain. Many craters, valleys and landslip features can be found within the rim alone. It is interesting that the west walls are much more heavily eroded than the east walls, the latter preserving at least a portion of their terraces.

At the center of the floor is what Goodacre aptly described as a "large massive triangular mountain," surrounded at its base by a number of subtle craterlets. The rest of the central floor is fairly peppered with other craterlets, first observed in 1897 by Percy Molesworth, who described them as "bowl-like depressions." Many of these are visible in my 250 mm (10-inch) telescope. Along the north wall can be found an interesting group of medium-sized craters, the largest of which, **Albategnius B**, has a craterlet of its own near its center (both Goodacre and Wilkins and Moore called this craterlet a central peak). A much larger structure designated **Klein** (formerly **Albategnius A**) intrudes into the southwest rim; its diameter is 32 km (20 miles). In many respects, Klein looks like a smaller version of Albategnius itself. There are many craterlets contained within its own rim, and it has its own central peak. Just outside the south wall of Albategnius is the barely recognizable walled plain **Parrot**, southeast of which runs a striking chain of at least eight overlapping craters – how many can you count?

Examine the landscape all around Albategnius and Hipparchus. If you are observant, you will no doubt notice many valleys and huge trenches cut through the rock; these are consequences of the Imbrium impact, as they all radiate from the approximate epicenter of the site of that great explosion. One of these clefts is actually visible inside Hipparchus, tangential to the southwest rim of Horrocks.

Figure 5.1 is a nice drawing of Albategnius and Hipparchus on the morning terminator, the observer/artist noting that she "was lucky to catch [Hipparchus] ... whole" as it rarely appears as anything more than fragments not obviously part of a single formation. Notice that the walls of Albategnius clearly rise much higher above the surrounding terrain than the walls of Hipparchus. How can we tell this just from the drawing? The floor of Albategnius is completely in deep shadow, the sunlight being blocked by the peaks along the eastern wall of the formation. But the floor of Hipparchus is brightly lit, indicating that the Sun has already cleared its east wall, which must therefore be much lower than that of its southerly neighbor.

Halley is a walled plain 34 km (21 miles) across with bright, high walls, broad and well-terraced on their inner slopes. Two craterlets break up the southern rim, giving the impression of a valley or pass. There is quite a lot of bright ejecta just outside the western ramparts, and many craterlets disturbing the whole rim. What Goodacre and others describe as a "distinct pass ... leading to Hipparchus" through the northwest wall is actually just a large craterlet located on the crest of the wall. The floor is fairly featureless, save for some very tiny craterlets, including one very near the center that was missed by some of the classic lunar mappers.

Airy is an attractive 35 km (22-mile) diameter crater that is part of an interesting chain of like-sized craters that curves south of Albategnius. Its walls show extensive terraces with an especially large landslip inside the west wall, and are well-peppered by craterlets. To the northwest is the long, impressive crater chain that carves a path towards Parrot.

Figure 5.1. Albategnius (top) and Hipparchus on the morning terminator. *Illustration by Sally Beaumont.*

La Caille is another beautiful crater, 56 km (35 miles) across and almost a perfect circle. The walls of La Caille appear very rugged because they are so broken up by secondary cratering. They are highest on the east side, where they are topped by the craterlet **La Caille D**. Two other perfect bowl craters crown the northwest and southwest rims. Abutting the east and northeast walls on the exterior are three interesting depressions. The first, just east of La Caille D, is **La Caille E**, a shallow crater almost bisected by a strong central ridge. North of this crater is the formation **Delaunay**, a rectangular-shaped crater that is almost certainly a double or multiple impact site. Delaunay is completely bisected by a high ridge that is even more impressive than the ridge that divides La Caille E. This ridge is almost surely the southeast wall of the more recent impact site. Southwest of Delaunay, touching the north wall of La Caille, is **La Caille M**, which also has a decidedly rectangular shape. If you use high power on your telescope to closely examine La Caille M, you may detect a strange series of bright scourings that run from southwest to northeast. The floor of La Caille itself is lava-smoothed, but contains many minuscule craterlets that require very high magnification to be observed, an indicator that this region of the Moon has received its fair share of relatively recent impacts.

Werner, 72 km (45 miles) across, is an extremely circular crater ring that is one of the freshest on the lunar surface. It has very finely terraced walls that peak as much as 4500 m (15 000 ft) above its floor. Since the Moon is a mere quarter of the Earth's diameter, that is like having an 18 000 m (60 000 ft) high mountain on Earth. Its central peak is no less majestic, relatively speaking, with a height of 1400 m (4500 ft); it is the highest of several well-defined hills on the interior floor. The terraces of the inner slopes spill far onto this floor, where they take the form of curved, concentric ridges. To test your telescope and your vision, search for the two craterlets just inside the northwest wall, one slightly smaller than the other.

Aliacensis is an 84 km (52-mile) wide crater with broad but eroded walls located in the Moon's southern highlands. Aliacensis has walls that are 3500 m (12 000 ft) above its floor on its east side, but as much as 5000 m (16 500 ft) above the floor on the west. The eastern wall is very serpentine, bowing outward at two places along its southeastern boundary. A weirdly distorted crater called **Aliacensis W**, crowns the south wall, and numerous alternating depressions and promontories can be seen along the western ramparts. The floor may at first glance seem to be devoid of any significant features, but you should be able to spot a small peak just north of center, and, more challenging still, a series of roughly parallel ridges on the northeast floor running from northwest to southeast.

The name **Theon** actually refers to a pair of bright, deep craters, curiously designated **Theon Senior** and **Theon Junior**, lying to the west of Delambre. As you may have already guessed, Theon Senior is larger than Theon Junior, but not by much – 18 km (11 miles) to 16 km (10 miles).

Alfraganus is a 19 km (12-mile) wide crater at the center of a modest system of rays. Alfraganus is basically a bowl crater whose walls are not terribly interesting, but its floor does have a mountain group consisting of two ridges at right angles to one another. At full moon, like most ray craters, Alfraganus is very bright.

Abulfeda, a 65 km (40-mile) wide walled plain, has complex walls but a rather featureless floor. The walls are 2750 m (9000 ft) high on the east, 3000 m (10 000 ft) on the west, with large landslips cutting diagonally along the northwest wall, and rows of summit craters topping both the north and south walls. The structure of the south wall is especially intriguing. The entire rampart is peppered with numerous craters which have obliterated any evidence of terracing which may have previously existed. The outer south wall comes to an apex where it adjoins an unusual depression that is not shown on the standard lunar charts. There is a sizable craterlet at this apex which marks the beginning of a remarkable crater-chain that extends for one 240 km (150 miles) to the southeast, towards the Altai Mountains. The inner south wall of Abulfeda is a very straight row of hills, many capped by craters as large as the whole hill. The inner north wall is also very linear, in an east–west direction, giving the floor of the crater a somewhat rectangular outline. Besides a few crater-pits and bright patches, the floor is dark and devoid of major features.

You will want to explore the Descartes region northeast of Abulfeda. **Descartes** is a 48 km (30-mile) wide impact site whose outline is difficult to trace, the north "wall" being all but gone. A few hills are all that mark the south boundary. Much more conspicuous is a large bowl crater that intrudes into the southwest rim of this formation. The most distinguishing features of Descartes are the series of ridges that curve across its floor, roughly concentric with the southeast and northwest walls. These ridges give Descartes the appearance of a wild animal's pawprint, pressed into the lunar soil. Figure 5.2 illustrates the appearance of Abulfeda and Descartes near sunrise. Notice how much work goes into an accurate drawing of even a "simple" region of the Moon.

Almanon is a walled plain just southeast of Abulfeda, but a little smaller at 45 km (28 miles) in diameter. The east walls appear to be much higher than the west walls, and in fact they are: 2000 m (6000 ft) as opposed to 1200 m (4000 ft). The south wall is very squared-off. Sizable, shallow depressions are scattered over Almanon's floor, which looks much like that of Abulfeda. A very large craterlet, **Almanon C**, is tangential to the northeast rim; it shares its east wall with a smaller craterlet that has a central ridge. Tangential to the outer south wall is **Almanon A**, a deep craterlet which

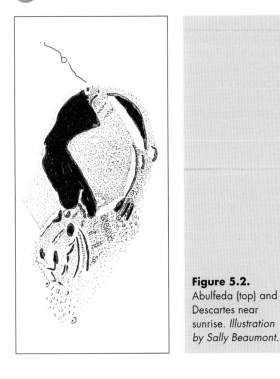

Figure 5.2.
Abulfeda (top) and
Descartes near
sunrise. *Illustration
by Sally Beaumont.*

rests atop the north wall of **Almanon B**, a much larger ring having a bright central peak.

Abenezra, an unusual 43 km (27-mile) wide crater, forms a trio with Azophi to its southeast and Abenezra C, a ruined ring to its west. The 4500 m (15 000 ft) high rim of Abenezra is razor-sharp and rippled; its southeast wall is shared with Azophi. The floor of Abenezra is fascinating – there is a central mountain from which radiates a series of clockwise-curving ridges that make the formation look like a pinwheel. A few ridges curve in a counterclockwise direction, toward the northeast. High magnifications will reward the patient observer with hours of enjoyment as the changing sunlight plays off these ridges at varying angles. **Abenezra C** is an ancient ring whose northeastern third is covered by Abenezra, which was obviously formed at a later date. It appears that the impact that created Abenezra threw debris onto the floor of Abenezra C in a weird pattern that mirrors the curved ridges on the east side of Abenezra.

The **Altai Mountains** are a well-formed mountain range that extends 505 km (315 miles) from the west wall of Piccolomini to the west side of the large crater Catharina, cut by four huge cross-faults. You will want to look for the Altai Mountains when the Moon is six days old; using low power, you will notice that this curved range is roughly concentric with the southwestern "shore" of Mare Nectaris. Planetary geologists believe that the Altai Mountains are the lone remnant of a much larger

impact basin that surrounded Nectaris, whose remaining rims were obliterated long ago by the numerous newer impacts in the region. It has been speculated that the Nectaris–Altai impact complex might have appeared at one time as a giant bull's-eye, with one impact basin centered inside the other. It is difficult to imagine such a formation today.

The Altai Mountains rise very steeply from plains to the east, averaging 1800 m (6000 ft) or so, with occasional higher peaks (two reach altitudes of 3500 m, 11 000 ft, and 4000 m, 13 000 ft). Notice how deep the shadows are at sunset, and contrast this appearance to how the mountains look at sunrise, when they appear as a great row of white, shining cliffs. The great crater Piccolomini stands guard at the southeastern tip of the range; this part of the mountains contains many lofty peaks. The mountains snake their way northwestward, interrupted in many places by craters large and small. They narrow considerably at their northern terminus near Catharina.

Theophilus is one of the most important walled plains on the Moon's surface, comparable to Copernicus in its stature. Like most of the large, highly detailed lunar craters, Theophilus is young and well-preserved; it forms a striking group with the nearly identically sized craters Cyrillus and Catharina, rimming the western wall of Mare Nectaris. At 103 km (64 miles) across with majestic walls rising to between 4250 and 5500 m (14 000 and 18 000 ft) above its surface, Theophilus displays all of the characteristics of a fresh impact site. It has perhaps the most impressive group of central mountains and ridges of any lunar crater – these catch the rays of the rising Sun well before the rest of the formation becomes visible. Look for the crater that has indented the northern group of peaks. The southern group of central mountains boasts several rugged ridges on its south slopes, separated by dramatically deep valleys. It is fun to watch the rapidly-shifting shadows in these valleys as the Sun rises or sets over the formation. Many tiny craterlets pock the surface at the foot of these central mountains. West of the central peaks you will see much rough ground, including numerous lower ridges and hills. The eastern floor, particularly toward the southeast wall, contains several low, dome-like hills.

The walls of Theophilus display almost too much detail to accurately sketch. The terracing of the very broad slopes of the inner west wall is extensive and complicated, with huge landslips visible everywhere. There is a generous depression on the southwest rim, which bows out where it overlaps Cyrillus, and a fine bowl crater situated on a broad ledge along the northwest rim. The inner eastern wall rises much more steeply from the floor than the opposite (west) wall, but it also has its share of terraces. A fine jagged rille zigzags its way from the crest of the northeast rim to the floor. Many hillocks dot the lower slopes of the inner north wall, and the outer portion of this rampart is accented by sharp ridges. The south walls are less perfectly contiguous than the rest of the ramparts, and contain a number of large rounded hills.

Cyrillus is a polygonal 90 km (55-mile) wide walled plain whose northeast wall is overlapped by the more recent impact site, Theophilus. The crest of the eastern wall of Cyrillus consists of two very straight segments that join each other at a sharp angle; the inside of this wall slopes very gradually down to the floor of the crater, broken at two levels by long landslips parallel to the outer wall segments. The west wall is cut by a long rille that meanders north and south, and higher up, on the northwest crest, are several wide depressions. Far inside the northeast wall is a very long, curved ridge. The region outside the rims of Cyrillus features many low rings and depressions, separated by a landscape that alternates ridges and gorges. A very broad, shallow valley connects the south wall of Cyrillus to its neighbor Catharina; you will need very low angles of illumination to see this valley.

The floor of Cyrillus is very rough and complicated, and there is much to see here. The most prominent feature is a group of three broad mountains between the center and the east wall which, under certain lighting conditions, may be mistaken for shallow depressions. This kind of optical illusion is common when observing the Moon – I have alternately seen craters as elevated topography and vice versa, even during the same observing session, an interesting phenomenon mentioned by many lunar observers I know. Although these mountains are massive, they are not nearly as lofty or sharply defined as those at the center of Theophilus, yet another indicator that Cyrillus is the older of the two. A large triangular valley separates the mountains. There are true craters on the floor, the largest being **Cyrillus A** in the southwest corner. Rocky ridges radiate in all directions from this crater. Look also for a sinuous rille that begins at an obvious volcanic vent some distance beyond the south wall of Cyrillus, and runs over that wall and down onto the southern half of the floor, where it zigzags its way across to a tiny craterlet just southwest of center. This rille, almost certainly a lava tube, looks much like Schröter's Valley. A much straighter rille is just south of center, crossing the floor from southwest to northeast; its length is almost half of the floor's breadth.

Catharina is the oldest crater of the trio it forms with Theophilus and Cyrillus. Catharina has badly eroded, very irregularly shaped walls that span a distance of some 90 km (55 miles). Several large craters overlap one another atop the northeast wall, having largely obscured this part of the rim The western rims of these craters join to form a single high ridge. Evidence of smaller destructive impacts may be detected all along the rest of the ramparts, which show no terracing on their inner slopes. Overlapping the northern third of Catharina's fairly smooth floor is the highly eroded ring known as **Catharina P**. The floor of this ring contains numerous crater pits and is cut by a rille running diagonally inside its southeast rim. **Catharina S** is a circular ghost crater near the south wall of the main crater. Unlike most ghost craters, it does appear to have a central peak, and overall its floor is atypically rough. Beyond the south wall is **Catharina C**, a miniature version of Catharina itself, a very similar chain of impacts having obliterated its eastern wall also!

Forming a pair with Capella, with which it shares its eastern wall, **Isidorus** is a 48 km (30-mile) diameter crater forming almost a perfect circle with steep walls and a dark, smooth floor. Isidorus has a nice bowl crater, **Isidorus A**, inside its west wall, and a tinier crater abutting it to the north. **Capella** is of similar size to Isidorus, but very different in appearance. It overlays its neighbor and is less depressed below the surrounding terrain, its walls broader and more gradually sloped. The walls of Capella are also very irregular, especially on their southeastern side, where a huge promontory juts out onto the floor. A distinct crater tops this promontory. The inner western slopes are dotted with much impact debris, forming tiny hills that brighten quickly with the first rays of sunlight. You will want to carefully observe Capella's large central mountain, which is capped by a summit crater that makes it look very much like a terrestrial volcano. Between Isidorus and Capella and adjoining their southern rims are two old, shallow, overlapping impact sites, the largest designated **Isidorus F**. Running northwest of Capella is an interesting chain of large, overlapping craters; from a craterlet topping the southeast rim a long, wide gorge heads southeast, past the crater **Gaudibert**.

Censorinus, a small but very bright crater, is located on a bluff overlooking the southeast "shore" of Mare Tranquillitatis. Only 5 km (3 miles) across, Censorinus is nevertheless an easy object in small telescopes, as it is one of the brightest objects on the Moon's surface.

Messier and **Messier A** are a pair of small, deep, bright craters on the western lava plains of Mare

Fecunditatis. Messier is a 14 km (9-mile) wide ellipse; the more circular Messier A has also been known historically as **W.H. Pickering**, after the American astronomer who thought that the changing appearance of this pair was caused by the formation of patches of hoarfrost. In reality, the changes are attributable to the changing angle of the Sun's rays as they strike the bright ejecta around these craters. The two craters seem to take on different geometrical shapes as the Sun rises and sets over them, sometimes looking very round, sometimes more elliptical, sometimes almost triangular. The most obvious feature of the ejecta is a pair of bright rays that diverge away from the craters towards the west. Goodacre called these rays "the comet tail," and they do indeed bear a striking resemblance to the that unrelated celestial phenomenon.

The contrast between Messier and Messier A is most apparent at first quarter. The ever-changing appearance of this pair was described by Pickering, who studied them as much as anyone:

> a series of periodic changes dependent on some unknown cause do occur, producing a series of most singular variations in the appearance of these objects. Thus sometimes one crater and sometimes the other appears the larger, while again both appear exactly the same size, even more remarkable changes take place in their shapes, especially in the case of [Messier] A.

You can see Messier and Messier A when the Moon is a three-day-old crescent. See if you can spot the companion crater to Messier A, which looks like a bright elongation of its western wall. Most of this companion appears to have been obliterated by Messier A itself, but many planetary scientists believe that all three craters, as well as the rays, resulted from a single, grazing impact of a large meteoroid with an incident angle of perhaps as low as 5°.

Gutenberg is a 72 km (45–mile) wide walled plain whose interior has been greatly modified by lava flows from Mare Fecunditatis to the east. Older lunar guidebooks call Gutenberg "irregular," but this is a misnomer: the walls of this crater form almost a perfect circle, though on the east side a 24 km (15-mile) wide crater known as **Gutenberg E** intersects the wall, giving it the appearance of a diamond ring. On the opposite side of the formation another secondary crater appears – this is **Gutenberg A**. This 16 km (10-mile) wide crater is very brilliant at full moon. To the south is the heavily eroded ring **Gutenberg C**, typical of many such formations in the region. The floor of Gutenberg encloses a very interesting semicircle of

mountains open on the east side; these are almost surely the remains of an ancient impact site, perhaps also incorporating mountains formed by the impact that created the principal crater.

The **Pyrenees Mountains** are a smallish range forming part of the eastern boundary of Mare Nectaris, beginning at Gutenberg and continuing southward past the crater Bohnenberger. The highest peak in the south attains an altitude of 2000 m (6500 ft); a much loftier peak may be spotted at the north end of the range, at 3500 m (12 000 ft). The slopes of the Pyrenees are very steep on their west side, where they rise abruptly from the lava beds of Mare Nectaris. Parts of the range consist of parallel high ridges, especially near Bohnenberger and Colombo. Look for the split in these ridges west of Colombo A, forming a valley with an isolated mountain that sits between the ridges.

Colombo, named for Christopher Columbus, is an 80 km (50-mile) walled plain that sits perched atop highlands that separate Mare Nectaris from Mare Fecunditatis. The older guidebooks compliment this formation as "fine and prominent," but I have always found that its resurfaced walls and poor contrast with the bright surrounding highland terrain make it somewhat difficult to identify . The many affronts to its walls do indeed give Colombo a most interesting lopsided shape, rather like a lemon. The inner slopes of the east ramparts are very broad and contain three concentric ridges which make it "appear triple," according to Goodacre. **Colombo A**, 40 km (25 miles) in diameter, is a more nearly circular ring with a very broad inner east wall that abuts the northwest wall of the main crater. Wilkins and Moore charted at least ten craterlets on the floor of Colombo – how many do you count?

Borda is a low-walled crater 42 km (26 miles) across with a very unique and irregular appearance. To me it looks like a frying pan. This illusion is created by a crater chain that juts from the northwest wall – the "handle" of the pan. Inside the "pan" are two "eggs" – one sunny-side up, the other down! Actually, these are a dome-like central hill and a shallow bowl crater on the floor. Just outside the southwest wall are two weirdly shaped impact sites, one lying over the other. See if you can spot the majestic 3500 m (11 000 ft) high mountain peak crowning the eastern rampart of Borda. The landscape around this crater is everywhere indented by sizable, shallow rings, separated by rolling hills. To the southeast lies one end of the **Vallis Snellius**, a very long and very wide valley bisected by its namesake crater.

Langrenus, together with Vendelinus, are a matched pair of very large walled plains that decorate the eastern "shore" of Mare Fecunditatis. Roughly 135 km (85 miles) across, Langrenus is elongated considerably in a north–south direction by its closeness to the Moon's east limb. If we could fly directly over this great impact site, we would of course see it as much rounder. No doubt, were Langrenus located nearer the center of the Moon's nearside, it would be a much more imposing object. Even so, it is impressive, its inner western ramparts being perhaps the most complexly and beautifully structured of any crater, an intricate intertwining of braided terraces and ridges. Its eastern slopes are probably similar, though their poor location makes them more difficult to see in great detail. The rim of Langrenus rises about 2750 m (9000 ft) above its floor, which contains a nice group of 1000 m (3300 ft) high central mountains cut diagonally by a dramatically deep canyon.

Besides these mountains, there are no other easy objects on the extremely bright floor of Langrenus – in my 250 mm (10-inch) reflector I can barely make out an old, buried crater ring due north of these mountains, but little else. Wilkins and Moore found some delicate rilles in the interior, but these must be exceedingly difficult objects. Perhaps the brightness of the floor is partly to blame for these difficulties. The south wall, punctuated by many rounded hills, gradually rises to meet the terrain beyond its crest. In contrast to the smooth mare basalts to the west, the landscape east of Langrenus is extraordinarily rugged, with many large craters and innumerable smaller ones. Looking toward the limb, it is occasionally possible to see several mountain ranges in profile – in 1948, the well-known British lunar and planetary observer Richard Baum discovered a ghostly bluish mountain range on the southeast limb.

Vendelinus is an extremely battered ring plain south of Langrenus, having an irregular, oblong outline. Located approximately halfway between the huge walled plains Langrenus and Petavius, Vendelinus is more than 160 km (100 miles) in diameter. Like its giant neighbors, it is foreshortened, but unlike most features subject to this limb effect, Vendelinus is not as elongated as you might expect. Why is this so? If you were able to look directly down upon this crater, you would notice that its true shape is elongated east to west, with squarish corners. This east–west elongation compensates for much of the north–south foreshortening, so that the apparent outline of this formation as seen from Earth is fairly round.

Vendelinus must be considerably older than Langrenus, as it has a very broken and worn appearance. Its walls are not high and sharp, but low and

chaotic. They are not terraced, but you can easily count many fresher craters marring the ramparts. The largest, **Lamé**, half as large as Vendelinus itself, overlaps the northeast wall. But Lamé is itself a victim of subsequent cratering – look for the chain of four sizable craters on its east wall. A huge depression abuts the north wall, which is completely broken down at that point. As with Langrenus, the region around Vendelinus, except on the mare-smoothed west side, is heavily cratered. Many large impact sites can be found to the north, south and east; the crater **Lohse** touches the north wall of Vendelinus, while the crater **Holden**, about 40 km (25 miles) wide, abuts the south rim.

Not surprisingly, the floor of Vendelinus is also pocked with large crater rings and a smattering of bowl craters. There are no central peaks, just a few low ridges that jut from the walls. Under very high magnifications and under extremely steady skies, using a 250 mm (10-inch) telescope I have glimpsed several very short gouges on the floor, all oriented north–south. These gouges were much more numerous in the great 40-inch (1 m) refractor of Yerkes Observatory, where they appeared to be of two different "families" – one set running from southwest to northeast, the other diverging in a northwest to southeast direction. The gashes appeared to converge at some ill-defined point near the north rim. Several classic observers reported glimpsing these gouges, which they described as being at the very limits of their instruments, and which they further assumed were parts of long, continuous rilles; but this is not the case – they are just broken segments, not connected to one another. Careful studies have shown that, when observing features at the threshold of resolution, the observer's brain sometimes has a tendency to mentally connect up features which the eye actually sees as disconnected fragments, "convincing" the observer that they are in fact seeing a continuous linear feature. It is suspected that the infamous "canals" of Giovanni Schiaparelli's and Percival Lowell's Mars drawings are another artifact of this optical illusion. Figure 5.3 is an exceptionally beautiful rendering of Vendelinus and Langrenus.

Petavius stands in stark contrast to the broken-down Vendelinus – a great walled plain of more than 160 km (100 miles) breadth, with rugged, magnificently terraced walls, rambling central mountains and several deep rilles crossing its interesting floor. Petavius truly deserves its reputation as one of the most breathtaking of all lunar formations. It is often written that Petavius is "nearly circular," appearing elliptical only by foreshortening. But that is not exactly true – while it is foreshortened to some extent, when viewed from above by the Lunar

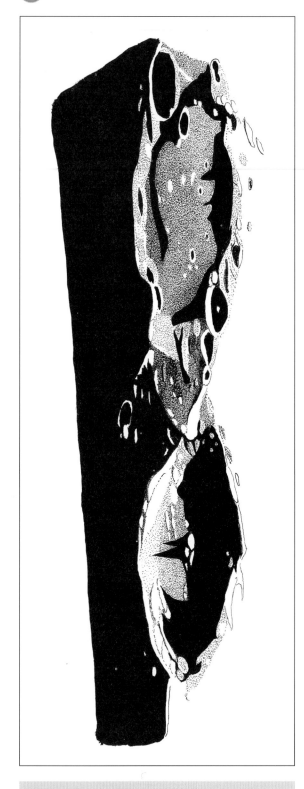

Figure 5.3. Vendelinus (top) and Langrenus; also shown are the smaller craters Holden (at the very top) and Lohse (on the lower right rim of Vendelinus). *Illustration by Colin Ebdon*.

Orbiter spacecraft, this crater was still quite elliptical, being longer from north to south than it is east to west. In fact, its floor when viewed from directly overhead has a distinctive pear shape to it.

It has also been written many times that Petavius would be more spectacular if it were located nearer the center of the nearside. While that is a matter of opinion rather than fact, I am not so sure I agree with that statement either – Petavius's placement so near the Moon's east limb only serves to heighten its dramatic appearance. Yet another myth associated with Petavius is the belief that it cannot be seen around full moon, but many observers have detected it when the Sun is overhead, and it is easily visible only one day after the Moon is full.

The walls of Petavius are truly massive affairs, rising 2000 m (7000 ft) above the floor on the east and higher still – 3500 m (11 000 ft) – on the west. They are extremely wide also, though this is not readily apparent because of the oblique angle from which we are forced to view the crater. The terracing along the inner slopes is often tortured, many offsets and depressions breaking up the symmetry. Some of the offsets create deep canyons within the walls themselves, and the play of light and shadow within them is absolutely fascinating to watch as the Sun rises over the formation. Wilkins and Moore, carefully studying the structure of Petavius's walls, concluded that the ramparts were once double, with only a trace remaining of the second wall. Just inside the west wall, running for much of its length, is a very fine rille – can you see it? Its darkness is startling – there is nothing else quite like it on the Moon's nearside. There are also ravines and canyons which cut through the walls longitudinally; a particularly deep canyon cuts through the southern wall, a little west of a small, deep crater that crowns the rim.

The floor of Petavius deserves your close attention, as it contains an interesting variety of details that appear in sharp relief at low Sun angles. Foremost are the central peaks, grouped in a 50 km^2 (20 square mile) area surrounding a bright central valley. These mountains appear to be connected to one another by a 1500 m (5000 ft) high ridge in the shape of the letter J (viewed with south "up"). There are three major peaks and eight or nine minor ones, attended by many smaller foothills at their base. To their southeast is the sharp, perfectly round crater designated **Petavius A**.

Connecting these mountains to the southwest wall of Petavius is a deep rille that probably formed when the intruding lavas cooled and contracted. This rille, discovered by Schröter in 1788, is an easy object even in small telescopes (it can be seen in a

50 mm (2-inch) refractor!) and is Petavius's trademark feature. It is 40 km (25 miles) long and 3 km (2 miles) wide. Look for another deep rille that runs from the north wall down into the central mountains, where it opens into the brightly lit valley. Other, more delicate rilles may be found running north and south along the eastern half of Petavius's floor. The floor itself is not of uniform coloration, with alternating dark and light patches. The geological experiments conducted by Apollo 17 suggest that the darker patches are older lava deposits. This finding surprised many planetary geologists, who had guessed that the dark patches were newer deposits.

Before leaving Petavius, make sure you take the time to explore the regions outside its walls, especially to the east, where you will find the chain of huge overlapping impact sites known as the **Palitzsch Valley**. It is really not a valley, but it does look like one because it is partially hidden from view by the east wall of Petavius, and is further distorted by foreshortening. It is in fact a series of four or five craters overlapping one another in a north–south line. If you look very closely, you can just make out the dilapidated rims of a sixth and a seventh crater just above the four or five prominent ones. Just beyond the northwest wall of Petavius is the fine crater **Wrottesley**, which has everything a well-preserved crater should have, including richly terraced, lofty 2500 m (8000 ft) high walls and a fine group of central peaks. Look also for the large, disturbed crater ring known as **Hase** which adjoins the southeast wall of Petavius, with its equally complex walls and rough, disorganized floor.

Furnerius is an impressive, 135 km (85-mile) wide formation with its 3500 m (11 500 ft) high walls, which you will find south of Petavius. Furnerius was not well imaged by the Lunar Orbiters, dispelling the incorrect notion held by many that there is no point in observing the Moon with Earth-based telescopes since spacecraft have completely "mapped it out." In fact, most planetary scientists would tell you that we still don't have a map of the Moon that is as accurate as they would like. We do know much about this great walled plain, however, thanks largely to the assiduous efforts of amateur observers. In the eyepiece Furnerius appears as a great oval, elongated north–south by its location near the east limb. In reality, it is slightly elongated in the opposite direction, east to west.

Furnerius is obviously very ancient – its walls are difficult to separate from the many ridges created all around them by subsequent impacts of meteoroids, there is no terracing to speak of along the inner slopes, and the whole interior is heavily cratered.

Nor is there a central peak or any mountains of appreciable height – just a few badly eroded hills here and there. The central floor is fairly smooth, but the east and west sides of the floor are, in contrast, quite rough and very heavily dimpled with craters of every size imaginable. This suggests that perhaps the lava flows inside this crater were not terribly extensive, filling in only the lowest (central) part, while leaving the outer margins relatively unscathed.

Look for a huge curved rille which cuts through the north floor, as well as the large crater **Furnerius B** to the southwest of it. Carefully examine the western floor under high powers and you will be rewarded with views of several short gouges, all oriented approximately north–south. They appear to trace their origin to an impact somewhere northeast of Furnerius, but where, exactly? The landscape outside the walls of Furnerius looks very much like the floor – many craters of all dimensions, the ground between them looking extremely battered and worn. Figure 5.4 is a dramatic rendering of Furnerius at sunset.

Snellius is a rather ugly, beat-up crater of 80 km (50 miles) breadth that looks as if it was used for target practice by some extra-terrestrial Air Force! Actually, in the years during and after World War II, scientists who studied the craters made by bombs finally discovered the details of how such explosions excavate debris from the target area and redistribute it through shock waves.

In the case of Snellius, the "bombs" were probably small pieces of asteroids, but the end result was the same – a heavily battered landscape pocked with craters of many different sizes. In fact, the damage is so extreme that this poor crater's rim is nothing more than a ring of craters, the original peaks having been knocked flat almost everywhere you look. Still, the walls manage to rise 2300 m (7500 ft) above the floor, which is likewise scarred. The most significant feature of the floor is a well-centered curving rille. But Snellius itself is not nearly as well known as the famous crater chain named after it – the **Snellius Valley**. This is a 240 km (150-mile) long furrow in the lunar landscape, roughly centered on its namesake crater which interrupts its progress; not coincidentally, we can trace the Snellius Valley back to the center of the Mare Nectaris impact basin.

The Snellius Valley, together with the very similar Rheita Valley to its southwest, are almost surely the scars of the Nectaris impact event. I have detected suggestions of still other such gouges in the lunar landscape radiating from Mare Nectaris under low power and very low Sun angles, which occur in the

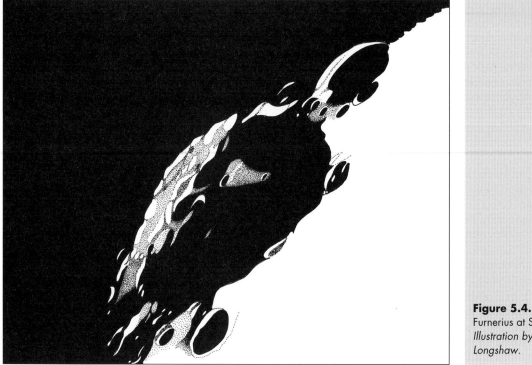

Figure 5.4.
Furnerius at Sunset.
*Illustration by Nigel
Longshaw.*

region a day or so following the full moon. It is probable that other, similar "valleys" existed all around Nectaris, but have been obscured by the countless later impacts that have obviously occurred here.

Stevinus, in sharp contrast to ugly Snellius to its north, is a most handsome crater – a marvelously perfect example of an impact site in every respect. Its beautifully terraced walls have exquisitely chiseled rims that rise as much as 3475 m (11 400 ft) above its equally well-preserved floor. The outline of Stevinus is an almost perfect circle with a diameter of 80 km (50 miles), though foreshortening naturally makes it look more elliptical than it really is. There are several levels to the terraces along the inner slopes, but all of these terraces are very concentric with the smooth outer rims. The central mountain range takes the form of an almost, but not quite, complete ring surrounding a very dark oval valley; closer inspection discloses that this range actually consists of a straight range on the east side of the valley and a series of individual lower hills on the west side.

Kästner, a great walled plain on the southwest shore of Mare Smythii, is a very challenging object to observe because it is virtually on the lunar limb, and is hence greatly distorted. Kästner appears as an extreme ellipse, and we should not expect to see it any other way. Like virtually all of the Moon's other

impact sites, it is of course really pretty circular, though it does have straight segments that make up its east and southeast walls. One interesting aspect of this crater is the chance positioning of four craters, all nearly identical in size, atop its ramparts at the northeast, southeast, southwest and northwest corners. If you look closely, you will be able to detect three more craters abutting or very close to the outer walls of Kästner, just slightly smaller than the four already mentioned. The floor is fairly featureless – this is not just because of the extreme foreshortening that afflicts Kästner and which would make any details very difficult to see, but also because the floor has been extensively resurfaced by lava flows.

Mare Smythii is a lava-filled basin at the extreme eastern edge of the Moon which can be seen only as a very elongated ellipse during favorable librations; at all other times Mare Smythii is nothing more than a slender darkened strip along the lunar limb. Wilkins and Moore found "ridges, craters, craterlets, hills and obscure rings" on its surface, but I have never had much luck in identifying much detail inside of this sea at all. On its western shore lie the craters **Gilbert** and Kästner (see the description above).

Wilhelm Humboldt is a huge ring-plain, 190 km (120 miles) across, that would be a spectacular object were it not on the Moon's east limb. Its walls are nothing less than precipitous, soaring 5000 m (16 000 ft) above its richly textured floor. Wilkins and Moore found the north wall to be double. Be sure to locate the startling chain of jagged mountain peaks that crosses Humboldt's floor from its center toward the northeast; these peaks are a row of perfect, bright triangles that cast long shadows upon the eastern wall at sunset. South of this range are two clumpy groups of mountains; to the southwest is a curved rille. A bright crater, **Wilhelm Humboldt N**, appears on the north floor. When the Moon presents a very favorable eastern libration, so that you can see the east side of Wilhelm Humboldt well, look for the 110 km (70-mile) wide crater to its southeast known as **Whitaker**, named for Ewen Whitaker, one of the lunar experts who helped construct the detailed maps that made the Apollo Moon landings possible.

Well-centered in the Moon's southern highlands you will discover an interesting pair of walled plains. **Stöfler** is a 135 km (85-mile) wide, smooth-floored depression whose circular borders are squashed somewhat by its closeness to the Moon's southern limb. **Faraday** is a dumbbell-shaped depression that overlaps Stöfler's southeast wall, apparently shoving much of the original wall material far onto Stöfler's floor, where it forms a badly broken ridge. Or is this ridge really the remnant of a separate impact within Stöfler? We don't really know for sure. The floor of Stöfler is otherwise remarkably smooth, except for mysterious bright streaks that are most likely part of Tycho's ray system.

The walls of Stöfler are worthy of careful study, despite the intrusion by Faraday. For one thing, they are different all round. From Faraday a valley curves along the northeast rampart, which then is broken by several craterlets. Next, almost due north is a wide gap in the wall, which is surprisingly straight on either side of the gap. Then you will come to a short section of the northwest wall that is shared with the crater Fernelius, and next to this section is the polygonal **Stöfler K**, a crater resting atop the northwest wall. One peak here attains a height of 3700 m (12 000 ft). Much of Stöfler's west wall appears double, thanks to a long, concentric landslip. The edges of the terraces here are very bright. Next you will see, interrupting the wall at the southwest corner, **Stöfler F**, perhaps the most perfect example of a bowl crater on the Moon's surface. Finally, the southern wall forms a bright, high ridge that crashes into **Faraday P**, a distorted ring which, together with **Faraday C**, abuts the southwestern rampart of Faraday itself. The whole region is a hodgepodge of warped rings and depressions, many competing with one another for space on the lunar surface.

To best observe Stöfler, let's take the advice of Thomas Gwynn Elger, who wrote more than a century ago:

> To view it and its surroundings at the most striking phase, it should be observed when the morning terminator lies a little [west] of the [east] wall. At this time the jagged, clean-cut shadows of the peaks on Faraday and the [east] border, the fine terraces, depressions and other features on the illuminated section of the gigantic rampart, and the smooth bluish-gray floor, combine to make a most beautiful telescopic picture.

Maurolycus is very similar in size and general shape to Stöfler. Elger had this to say about it: "This unquestionably ranks as one of the grandest walled plains on the Moon's visible surface, and when viewed under a low sun presents a spectacle which is not easily effaced from the mind." Elger got a little carried away, claiming that Maurolycus measured "fully 150 miles [240 km]" across; in truth its width is more like 109 km (68 miles). Nevertheless, it is a wonderfully complicated crater, and definitely an object you will want to spend many evenings observing. Elger was correct in stating that Maurolycus "includes ring plains, craters, crater rows and valleys – in short, almost every type of lunar formation."

As with Stöfler, the walls of Maurolycus exhibit a rich variety of features, depending where you look. A very dramatic, high, razor-sharp ridge characterizes the southeast wall; it breaks off abruptly at its northern terminus, before it can join with the rest of the wall. This ridge is perhaps 4250 m (14 000 ft) or higher. Below it you will see several ravines leading to the floor of Maurolycus. The extremely broad east wall zigzags a short distance before being interrupted by a tiny, bright summit crater, then continues as a graceful arc to form the north wall. This part of the rampart is even broader than the east wall, and slopes very gradually down to the level of the surrounding terrain, making it difficult to tell that it is a wall at all. The northwest wall is topped by the crater **Maurolycus F**, then come two smaller craterlets marking the west point; and finally, the wall takes the form of a very bright and rugged ridge that continues all the way to the south, where it suffers one final affront, by the crater **Maurolycus A**. In contrast to the east and north walls, the west and south walls of Maurolycus are rather narrow, with no terracing.

Several craters adorn the floor of Maurolycus, which also features a complex group of central mountains having a particularly bright ridge on the east side of the range. About this range, Elger said, "The central mountain is of great altitude, its loftiest peaks standing out amid the shadow long before a ray of sunlight has reached the lower slopes of the walls." Look for chains of sizable craters that overlap one another on the floor of Maurolycus – there are at least three of these features north and west of the central mountain range. There are no long rilles inside this walled plain, but with high magnifications you may be able to detect several shorter gouges. Also, take notice of the many ancient rings in the region immediately surrounding Maurolycus – in fact, Maurolycus overlies one of the these old rings, and it is easy to deduce several generations of impact events in this region. Figure 5.5 is a drawing of Maurolycus as observed on the terminator.

Baco, a 65 km (40-mile) wide ring plain near the Moon's south pole, is characterized by its lofty (4250 m, 14 000 ft) rim and its smooth, almost featureless floor. The floor of Baco does contain what Wilkins and Moore call a "low central mountain," but this extremely elusive object is no more than a modest hill. There is much evidence of lava flooding, which probably occurred through a low part of the southeastern wall. The whole landscape around Baco is rather interesting – many medium-sized craters with lava-flooded floors superimposed upon a dark lava plateau, but with few small craterlets to be found anywhere.

Lindenau is a 56 km (35-mile) diameter ring plain southwest of Piccolomini, situated inside a large ancient impact site that is now barely discernible. Lindenau is graced by 2600 m (8600 ft) high walls to the east, and 3700 m (12 000 ft) high

Figure 5.5. Maurolycus on the terminator. The crater in deep shadow to the left is Barocius. *Illustration by Sally Beaumont.*

ramparts to the west and northwest. Lindenau is a fresh crater with cleanly outlined rims and splashes of bright ejecta outside its east and west walls. The west wall is very broad and complexly terraced, with no fewer than four wide gorges indenting the terraces. On the west side, these terraces become ridges that radiate across the floor toward a group of low central mountains.

Piccolomini is a wonderful walled plain 90 km (56 miles) across, south of Fracastorius. The walls of Piccolomini tower 4500 m (15 000 ft) above its floor, which shows four very bright pyramidal central peaks. At low power the outline of this large crater is fairly round, but higher magnifications reveal many twists and turns in the crest of the ramparts, reflected in the many rows of terraces along their inner slopes. The terraces of Piccolomini are a magnificent sight at sunrise. On the east side the terraces are particularly well organized, maintaining an orderly alignment of three separate levels of landslips that are almost perfectly contiguous. The west slopes are more chaotic, with a deep ravine cutting through the southwest section. Look just beyond the northwest wall of Piccolomini and you will see the beginning of the Altai Mountains, which carry on to the northwest. Directly north of the main crater is a most fascinating grouping of large secondary craters associated with Piccolomini. They are clustered around the rims of **Piccolomini F**, which is all but rendered invisible by its many uninvited guests. Moving clockwise, from the upper right (northeast) side of this grouping, we encounter **Piccolomini E, D, M, C, A,** and **B.**

The south wall shows evidence of lava flows that have spilled down its slopes onto the floor below, and indeed the floor of Piccolomini is greatly smoothed over by these lavas, displaying only a few insignificant crater pits around the majestic central mountains. Look for two or three gaps in the south wall (two are easy, the third is more challenging), and see if you can detect the channels that flow into them from outside the crater's walls, to the south. Under high power, these channels take on the appearance of crater chains.

Fracastorius, a large bay at the southern end of Mare Nectaris, is actually an old walled plain whose north wall has been destroyed by the Nectaris lava flows. Partially destroyed crater rings like Fracastorius give planetary scientists invaluable insights into the Moon's geologic history, for they allow them to surmise past events that happened from tens of millions to billions of years ago. For example, just by looking at Fracastorius we can ascertain that the basaltic lavas that flooded the Mare Nectaris impact basin must have flowed southward, at least over the southern half of that basin (it is possible that the flows erupted from the middle of the basin and traveled outward in all directions).

Fracastorius is about 100 km (60 miles) wide, and as such is probably the finest example of a partly ruined ring on the Moon's surface. A few low hills are all that's left of the north wall, and you will have to catch these under a very low angle of solar illumination to see them at all. Look for tiny craters marking the rise of two very tenuous ridges from the solidified lava pools at the northeast and northwest "entrances" to the bay, gradually increasing their dimensions on both sides to broad, rugged bluffs disturbed everywhere by depressions and ravines. The slopes of the east wall are fairly symmetric, centered around a thin, bright ridge. The south wall is very broken and eroded – it is considerably flatter and broader than the east or west walls.

On its southeast crest is a bowl crater, **Fracastorius N**, with a key-shaped depression to its northwest; the southeast crest of the south wall is capped by **Fracastorius A**, a crater lying at the head of a short chain of three or four like-sized impact sites. To its northwest, disturbing the southwest wall of Fracastorius, is the much larger crater **Fracastorius D**, a teardrop-shaped depression produced by another series of small impact events. To its north lies **Fracastorius H**, a huge, rectangular depression whose right side is one of the few fragments of the main crater's west wall that we can still see. A shallow crater chain that is usually mapped as a wide rille runs northwest from this formation. **Fracastorius E** is another bowl crater above this, atop the northwest wall, which thereafter takes the form of a rocky ridge, gradually declining in altitude until it reaches the bay's floor, mimicking the ridge on the opposite (eastern) side of the formation.

Despite classic lunar observing guidebooks that claim this lava-flooded bay contains a "large amount of detail," the floor of Fracastorius lacks any major features that will impress in a small telescope. Having said that, it is true that highly ambitious amateur observers using large telescopes have successfully charted many small objects here. For example, a very detailed map appears on page 156 of Wilkins and Moore, showing good numbers of craterlets and clefts. These are all very inconspicuous, and even in sizable telescopes they require excellent sky steadiness if they are to be detected. You should be able to make out the bare remnants of a central peak – it is easy in my 250 mm (10-inch) reflector. Obviously, most of this peak now lies below the flooding lavas. You will also probably

notice a few white patches here and there around the interior of Fracastorius – these are tiny ejecta blankets that betray the locations of many of the craterlets. It is a common occurrence when observing the Moon that you can see a craterlet's white ejecta collar but not the crater-pit itself.

Before departing Fracastorius, use a low-power eyepiece to explore the surrounding terrain. To the north is the very smooth and featureless basin of Mare Nectaris. But to the east, south and west of Fracastorius there are moderately cratered highlands. At low power, can you see two chains of craters diverging from the south wall of Fracastorius? They look like a large letter V, and are just two of many such crater chains in the region.

Mare Nectaris is a dark, oval, lava-flooded impact basin that is picturesquely termed the Sea of Nectar. You should think of Mare Nectaris as a big crater 290 km (180 miles) across, created by a huge impact 3.92 billion years ago. Still, it is one of the smallest of the lunar "seas." To planetary geologists this formation is highly significant, for the impact creating it marks the beginning of the so-called Nectarian Era, during which time the lunar highlands formed. The Apollo missions brought back many highland rocks, whose ages tended to be around 3.9 billion years old, a little younger than if they had formed from the solidification of the Moon's magma ocean, 4.2 billion years ago. Scientists studying these complex rocks have concluded that they are more likely the result of massive impact shocks caused by a heavy meteoric bombardment that began with the Mare Nectaris impact event and ended around the time of the Imbrium impact 3.85 billion years ago. These massive impact shocks pulverized the lunar rock to great depths, and refigured the lunar surface.

Like all mare regions, Mare Nectaris consists of a smooth lava-flooded plain surrounded by highlands and large craters. You will see, at low and medium magnifications, a few wrinkle ridges on its western shores, and also some ghost craters – most notably **Daguerre** – along the north shores, but little else. Look also for the several bright rays that crisscross the eastern floor; these are from different impact sites located far outside Mare Nectaris. Careful studies of the floor have disclosed the existence of numerous fresh craterlets, but most of these are tiny and difficult to see excerpt with larger instruments and higher magnifications. The lone exception is the 16 km (10-mile) wide craterlet **Rosse** in the middle of the southern floor of Nectaris, an easy object even in a 75 mm (3-inch) refractor.

I have already described most of the grand walled plains and mountains that encircle Mare Nectaris and make up its borders. Gaudibert,

Capella and Isidorus crown the heavily cratered highlands above the northern shore of Nectaris, while the magisterial trio of Theophilus, Cyrillus, and Catharina are concentric with much of its western edge. **Beaumont** lies along the southwest shore, while Fracastorius, just discussed, is the clear highlight of the southern shore. The eastern boundary of Mare Nectaris is marked by a conglomerate of smaller craters and the Pyrenees Mountains. Here you will also find the interesting 35 km (22-mile) wide crater named **Bohnenberger**, with its very circular, sharp rims and intricate series of rilles snaking across its rough and varied floor. The most prominent of these rilles winds its way from the north wall to a group of small, bright central hills, while two other, shorter, rilles cross each other on the southwest floor, making a letter X. Look also for a long, bright white ridge running due north from Bohnenberger, and two contrasting secondary craters due south of the main crater, the sharp little bowl crater **Bohnenberger G**, and the mostly ruined ring plain **Bohnenberger A**.

Reichenbach, northwest of Stevinus, is a highly distorted, 50 km (30-mile) wide depression with lofty walls that are 3700 m (12 000 ft) or so above its dark and featureless floor. Rendered elliptical in appearance by its proximity to the Moon's southeast limb, Reichenbach's walls are heavily cratered, a condition typical of the whole region's topography. A huge rectangular notch in the southeast wall further deforms its outline. This notch is largely an illusion created by two bowl craters, known as **Reichenbach F** and **J**, which sit atop the wall on either side of the notch feature. **Reichenbach L** similarly disturbs the northwest wall. No terracing is detectable inside the walls, but the many large and small impacts that have obviously reshaped this formation would have destroyed any distinct ledges long ago. Look for two large, overlapping rings just north of Reichenbach – these are **Reichenbach A** and **B**.

The **Rheita Valley**, one of the finest examples of a lunar valley, is actually the scar of a secondary impact from the impact event that formed the Mare Nectaris basin. Some 185 km (115 miles) long and 25 km (15 miles) wide, the Rheita Valley is named for the 68 km (42-mile) diameter crater that lies at its northern reaches, from where it carves its way towards the Moon's far southwest limb. **Rheita** itself has very sharply-defined crests, thanks to its lofty, 4250 m (14 000 ft) high walls. The great valley appears to commence in a shallow bowl abutting the northwest wall of Rheita, heading south across the crater **Young**, then along the eastern boundary of the smaller **Mallet**, finally ending up west of

Reimarus. All along its length, the Rheita Valley is widened by roundish depressions, suggesting it is a crater chain, likely formed by unimaginably massive blocks of debris excavated by the Nectaris impact, sent skipping or tumbling across the lunar surface with enormous force. The valley looks like a slightly bent finger, pointing up at Mare Nectaris, its likely source. Look for little occasional cross-ridges all along the length of the Rheita Valley – these are almost surely the rims of the smallish craters that are chained together to form the valley.

Forming a conjoined pair with Fabricius, **Metius** is an impressive, 80 km (50-mile) wide crater with broad, massive walls that rise to a maximum height of 4000 m (13 000 ft). Both craters are interesting in their own way. Metius is located just southwest of the crater Rheita and west of one of the deepest parts of the associated Rheita Valley. The ground between Rheita and Metius is very rugged, with a wide splash of bright ejecta crossing the Rheita Valley longitudinally. Centered on this white splash is a fine rille that runs southwest to the rim of Metius, interrupted by a tiny craterlet. The rille crosses the northeast rim of Metius, appearing to spill down the slope of its inner wall, terminating at the crater **Metius B**, the largest object on the floor of the main crater.

Look also for **Metius G**, a shallow bowl crater on the east rim, which has an even tinier craterlet atop is southwest rim. **Metius F** is an eye-shaped crater on the pointy north rim of the main crater. The southwest wall of Metius, where it merges with the northeast wall of Fabricius, is straight, a condition commonly found in crater pairs. The inner walls do not show much terracing, but there are a few large landslips, especially inside the north and northeast walls. The floor of Metius has several shallow craterlets and a group of hills near the center that look very much like a question mark.

Fabricius is 90 km (55 miles) wide, with broadened, ropy walls that reach a height of 2950 m (9700 ft). The floor of Fabricius is most interesting, highlighted by two mountain ranges that show complex structure. The larger range, to the northwest of center, is cut by a series of valleys running northwest–southeast – I count at least six of them. The lower, less massive mountain range runs north–south along the crater's central axis. A very fine floor rille appears to connect the southern ends of the two ranges. Many, still lower ridges run radially around the inside walls of Fabricius; these are especially prominent inside the north and east slopes.

Look for three rilles which originate from the south rim of Fabricius. One crosses the large, inverted teardrop-shaped depression southeast of

Fabricius known as **Fabricius A**. The middle rille actually begins inside Fabricius, at the base of the lower central mountain range, and heads southward across the floor, up and over the south rampart, and across the outlying terrain to the large bowl crater **Fabricius K**. This rille throws off a branch to the southwest before it leaves the interior floor of Fabricius. This branch, after winding its way narrowly through the southwest wall of the main crater, suddenly widens once outside the rampart to become the third rille, cutting a very broad swath across the floor of the ruined formation **Janssen**. It is by far the longest and widest of the three rilles radiating from Fabricius.

The whole area around Fabricius is fascinating. For example, the teardrop known as Fabricius A, mentioned above, has a floor that is much darker than anything around it; you should be able to make out a series of rilles on its floor which appear to diverge from a bowl crater on its south rim designated **Fabricius J**. These divergent rilles, which alternate with ridges between them, give Fabricius A the look of a scallop shell. There is another unusual feature about Fabricius itself which you should take the time to study carefully – the outer slopes of its western walls extend outward for about 30 km (20 miles) in the form of ridges that are very concentric with the western rim. There is some evidence of this same "ripple" effect outside the walls on the east side of the formation, but the presence of Metius and other craters, including Fabricius A, interferes with attempts to verify these structures. Figure 5.6 is a dramatic drawing of Fabricius and Metius.

Fraunhofer is a 1000 m (3000 ft) deep, 50 km (30-mile) wide crater southwest of Furnerius, very near the Moon's limb. This and the remaining objects described in this chapter are all very much foreshortened by their locations near the southeastern limb or the Moon's south pole. Fraunhofer does not show a tremendous amount of detail; its dark floor has evidently been smoothed by lavas spilling inside it. The whole region between Fraunhofer and Mare Australe is very dark, even though many craters intervene. Most of these craters are lava-filled, and, not surprisingly, there are many ruined or partly ruined formations and ghost craters in the neighborhood. Not much is left of the north or west walls of Fraunhofer, as several sizable craters are found here, the largest of which is designated **Fraunhofer V**. Next to it is the bright bowl crater **Fraunhofer G**. The south wall is even more ruined, appearing to be almost level with the ground outside the crater; this is probably where the filling lavas intruded into Fraunhofer's interior.

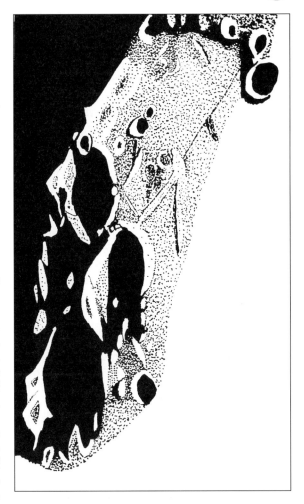

Figure 5.6. The craters Metius (bottom left, in deep shadow) and Fabricius (to its upper right). The largest of the small craters at top right is Lockyer. *Illustration by Sally Beaumont.*

Steinheil is another crater filled with solidified lava, its floor showing no more than a few low ridges and a couple of minute crater-pits. Steinheil is 72 km (45 miles) across, lying atop an older, more heavily eroded ring plain to its east known as **Watt**. The walls of Steinheil rise above its floor by as much as 3500 m (11 000 ft), and are finely terraced all the way round, but especially on the west. Look for a very elongated depression tucked inside the east wall, which is a little deformed where it overlies the west wall of Watt. In two places, the northwest rim bends inward, creating a meandering canyon just inside.

Vlacq is a 90 km (56-mile) walled plain with a bright, double-peaked central mountain and would be regarded as a much more impressive formation if it were not located so close to the Moon's southeast limb. Its walls rise to peaks that are 2500 m (8000 ft) on its east side, 3000 m (10 000 ft) on the west. Vlacq very nearly adjoins **Rosenberger**, a dark, lava-filled ring of about equal size to its southeast. Goodacre observed that "very few of these larger formations have received that telescopic scrutiny they deserve."

Just outside the northeast wall of Vlacq is the 40 km (25-mile) wide crater called **Rosenberger C**, which sports a bright little central mountain ridge. Just beyond the west wall of Vlacq you will find the interesting formation labeled **Hommel A**, a dark, irregular ring largely submerged in dark basalts, which itself encloses a smaller, much rounder crater, **Hommel V**, also filled to the brim with solidified lava. Two weird, distorted rings, known as **Hommel B** and **Q**, are a short distance beyond Vlacq's southwest wall, where they overlap a third old ring, **Vlacq G**.

Zach, at 74 km (46 miles) across and with walls that soar 4000 m (13 000 ft) above its smooth, dark floor, would be quite a prominent object were it not for its location very near the Moon's south pole, and amateur lunar observers rarely pay much attention to it. It lies in a remarkable region characterized by numerous large walled plains, many of which overlap or abut one another. You should be able to make out a small central peak and a couple of relatively fresh craterlets south of it, but little else on the floor of Zach. Its walls are very symmetrical, and probably nicely terraced, though the oblique angle at which we are forced to view this crater means that only a hint of terracing is visible.

Zach is attended by several sizable craters that either approach its rims or intrude upon them. Lying atop the northwest rampart is **Zach F**; it has a rocky protrusion from its southeast wall and a tiny central peak. **Zach D** is a crater of equal size that just about, but not quite, touches the southeast wall of the main crater. It has a couple of small craters on its floor, but no other detail of consequence. Squashed against the south wall of Zach is the oblong **Zach A**; less deformed is **Zach B**, just to its south. Look for the small group of central hills inside Zach B. You will want to take your time to explore the region around Zach, especially to the north, where you will find seemingly endless rows of shallow, lava-filled crater rings. One of the most remarkable of these groupings is the trio of **Lilius**, **Lilius A**, and **Jacobi**. Its large, bright, perfectly centered mountain peak makes Lilius is easy to spot; a true test object is the tiny rille between this big mountain and the northwest wall. Jacobi has no central peak, but five shallow bowl craters of almost identical size grace its floor.

Curtius, a mountain-walled depression of 80 km (50 miles) diameter situated not far from the south pole, is known for its surprisingly complex walls. It is quite a challenge to describe the walls of Curtius, for they show much variety. One thing is for certain – they are, in the words of Goodacre, "among the highest on the Moon's surface." Indeed, you will be impressed by an obviously massive mountain capping the northwest rampart which Elger quotes as having an altitude of 22 000 ft (6750 m)! The whole northwest wall is extremely rugged-looking, reminding the observer of the Earthly Himalayas. The north wall is no less spellbinding, cut by an enormous cigar-shaped canyon that runs west–east. A second huge ravine, almost as large, intrudes into the north wall from the northwest at an oblique angle. Other, lesser, gorges may be found all around the ramparts, the largest almost atop the west wall, running north–south.

The inner slopes of Curtius are as varied as the outer walls. A medium-sized crater, **Curtius E**, interlopes far onto the floor from the inner slope of the east wall. The inner slopes of the southeast wall appear broken down, but the inner slopes of the south and southwest walls display "gigantic terraces," in the words of Wilkins and Moore. The northwest inner slopes, if you can call them that, are a tortured jumble of very rugged mountains and lesser foothills. At this point there is a very oddly shaped gorge that drops precipitously to the floor below, and looks as if it was formed by multiple impacts, like a semi-vertical crater chain. You will find more gigantic terraces inside the north walls.

The floor of Curtius may be divided into two halves – the north half, which is very rough, heavily cratered and cut by trenches, and the much smoother, less interesting south half. Approximately in the center is a shallow crater; to its west is a rille, discovered by the lunar cartographer Ewen Whitaker, which cuts a path through the northwest wall. These features are surrounded on the north half of the floor by tiny crater pits in numbers almost too great to be accurately counted, much less mapped.

Malapert is noteworthy only because it is situated atop the Moon's south pole, and requires a very favorable libration in latitude for it to be observed. Even then, details within it are sketchy at best. It has high walls and is surrounded by several similar, but equally obscured, formations.

Webb, a bright, somewhat rectangular crater 23 km (14 miles) across, is located on the eastern margins of Mare Fecunditatis. From its northwest wall radiate several very long, wide rilles, one of which terminates in a remarkable crater chain. To the east is a long, high ridge that often appears as a white streak.

Chapter 6

Lunar Features – Northwest Quadrant

Murchison, a 56 km (35-mile) wide ringed plain, has been almost completely obliterated by erosion. Its west walls seem detached; the east walls are low and also eroded, but fairly bright and easy to make out even in small apertures. Murchison has often been compared to the walled plain Julius Caesar, described in the previous chapter. Only the northeast wall is intact. Southeast of Murchison is **Murchison A**, a very deep hole in the lunar surface, only 8 km (5 miles) wide but fully 1000 m (3000 ft) deep. The Dresden surveyor and lunar observer Wilhelm Gotthelf Lohrmann (1796–1840) called this crater **Chladni**, and it is still designated as such on many lunar maps. Murchison A is situated in an interesting terrain, on a bluff surrounded by a number of isolated mountain masses and craterlets. To the southwest is a shallow ring dubbed **Murchison E**, first reported by the American astronomer Edward Singleton Holden after he observed it in 1891 with the great 36-inch (0.9 m) refractor of Lick Observatory on Mt Hamilton, California.

Pallas, just west of Murchison, is about 50 km (30 miles) across and, like its neighbor, another great example of an eroded walled plain. There are many breaks in the walls, especially to the east, where the two features abut, and to the north. The west walls are broadly sloping to the interior floor and very bright; the east walls are narrower and darker. The south walls gradually merge with a smaller ruined ring named **Pallas B**, open on its south side. In between is a craterlet, **Pallas H**. The northwest wall is capped by a bright craterlet known as **Pallas A**. The floor of Pallas contains a typical central mountain mass.

Schröter, named in honor of the German amateur astronomer and selenographer Johann Schröter (1745–1816), is a ruined ring plain a little over 30 km (about 20 miles) wide. It looks rather like a pair of parentheses, as the walls are eroded to the north and south, leaving only the curves of the east and west walls to distinguish this formation from the inundating lava fields around it. Each half-wall is crested by a craterlet in its middle. The interior is, not at all surprisingly, virtually featureless save for a small craterlet just south of center, called **Schröter A**, from which William H. Pickering thought he saw an ejection of steam emanate.

Bode, easy to see even when the Moon is full (and that can't be said about too many craters!), is 18 km (11 miles) in diameter with very bright walls which rise 1500 m (5000 ft) above the interior. A mountain ridge runs along the crater floor to the north wall. There are quite a few craterlets along both the north and south boundaries. Just over 70 km (about 45 miles) to the northeast is a very conspicuous crater called **Bode A**, 10 km (6 miles) wide, which itself is pocked by a shallow ring, known as **Bode S**, on its northeast edge, through which runs a delicate rille. To the west of Bode A, past a shallow ring, is the bright crater **Bode B**, about 5 km (3 miles) across. **Bode D** is located southwest of the main crater, situated on a broad plateau with sloped sides; at the bottom of the western slope of this plateau is a 55 km (35-mile) long valley, curving to the southeast where it eventually disappears.

Look carefully between Bode B and D and you will see the ruins of an old crater ring which is broken on its south end, itself containing a number of craterlets that are a good test for small telescopes. Southwest of Bode D are two more ruined crater rings whose walls are very broken. Bode has a small ray system of its own. To the north of Bode lies a large but fairly shallow lowland area that is oval-shaped, the eastern side of which gradually merges into an extension of the Apennines.

Gambart is a walled plain nearly filled to its brim with solidified lava, giving it a very smooth and almost featureless floor. The careful observer will notice a bright patch on the southern floor. There is a wide pass in its southwest wall, and the much more symmetrical, smaller crater **Gambart A** lies due west beyond a dark lava plain. **Gambart B** and **C** are bowl craters to the east.

Sinus Aestuum, the Seething Bay, also known as the Bay of Billows, is a large lava plain to the east of Copernicus and Stadius and to the southeast of Eratosthenes. The terminus of the Apennines, a beautiful range of mountains pocked with deep craters, forms its northern border. This is the location of the unusual ghost crater Stadius.

Eratosthenes was called by Wilkins and Moore a "grand object," and rightfully so; 61 km (38 miles) across, it is found at the end of the Apennines, the magnificent chain of mountains that forms the northern border of Sinus Aestuum. If it weren't for nearby Copernicus, Eratosthenes would easily be considered the most magnificent object in its region, and there is so much detail in and around this walled plain that it could take the patient lunar observer many, many nights to note and draw all there is to see here. The east wall, about 3000 m (10 000 ft) high, is almost a perfect semicircle, but the 5000 m (16 000 ft) high west wall undulates wildly in great waves, and interestingly, the terraces here also undulate, often giving the wall a double appearance. Eratosthenes is a bit of a puzzle because, though it is obviously a well-formed and probably young, fresh crater, unlike most of its kind it is not easy to locate around full moon. The overwhelmingly bright ray system of Copernicus is partly to blame for Eratosthenes' disappearing act. At lower Sun angles, however, it is a most striking object indeed.

William H. Pickering depicted a series of canals and oases on the floor of Eratosthenes similar to the structures that Percival Lowell thought he saw on the planet Mars – he even reported his findings in the *Harvard College Observatory Annals*, Volume 53! Pickering, in true Lowellian fashion, assumed that these features belied the growth of vegetation, but of course we know that no such thing is possible on the harsh and barren lunar surface. But while there may be no plant life, there is still much to see on this crater's floor.

The central mountains of Eratosthenes take on a triangular shape, surrounding a crater-like depression. Deep ravines cut through the north slopes of these mountains. The terraces of the walls' inner slopes continue down to the floor, which is also covered with many hillocks, but few craterlets. Most of the craterlets that do exist are found on the south floor. The classic selenographers noted the presence

of several "submerged rings" on the crater floor, but I have never been able to see any, except for one northwest of the central mountains rather near the rim; it is a very challenging object. Mostly, the floor of this walled plain is dominated by impact melt material.

The crater's ejecta blanket is quite complex and subtly detailed. From the southwest side of Eratosthenes runs an impressive range of mountains which continue due south to Stadius. These mountains look very much like a corkscrew or a hand-drill bit. They do not come to sharp peaks, but are very flat on top, making them more of a chain of plateaus or cliffs rather than proper mountains. Many whitish streaks oriented east–west cross the lava fields to the south and east of the main walled plain. A short rille snakes northward from the north rim, and the Apennine range begins just outside the northeast rim.

Stadius is a remarkable and unique-looking walled plain, though one hardly knows where to look to find the walls themselves! This is because Stadius is perhaps the best-known example of a so-called ghost-crater – a crater whose walls have been so completely eroded by lava flows that one can barely make out its shape. The walls are best preserved to the northeast, at the foothills of the plateau ridge that runs southward from Eratosthenes. There is virtually no wall to the north, but a dark hint of a submerged rim appears in its place. A couple of low ridges suggest the presence of a southern rim. Because it lacks high walls, it is best to look for Stadius at low Sun angles, when relief shading is greatest. This happens when the Moon is 8 days old (right after first quarter) and again at 21–22 days into the lunation.

Perhaps the most striking feature of this ghost crater is the very great number of fresh craterlets that have dimpled the otherwise pristine mare material on its 65 km (40-mile) wide floor. Early selenographers tried to outdo one another in seeing how many of these craterlets they could count and draw; often, they had a tendency to exaggerate the number. Planetary geologists insist that careful studies show the interior of Stadius to contain no more of these small craters than the surrounding Sinus Aestuum, but those claims are hard to believe – anyone who has ever observed this formation will insist that Stadius is hoarding all of the region's craterlets to itself! They are so numerous that in several instances they share rims, bumping into one another. There are, however, no central mountains or even hills of any consequence, at least none that protrude above the lava flooding.

Northwest of the main crater curves **Rima Stadius I**, which is not really a rille but an impressive chain of craterlets that, like its brethren inside Stadius, is likely the result of secondary impacts from the catastrophic impact that produced nearby Copernicus. This crater chain continues all the way to the southern shore of Mare Imbrium. Another very nice chain of craterlets forms a semicircle arcs southeast of the main crater. Due north of Stadius and just west of Eratosthenes are two large mountain masses.

Copernicus – one could write an entire chapter on this great impact site which dominates the northwest quadrant! Without a doubt the finest walled plain on the lunar surface, Copernicus is 90 km (56 miles) across with amazingly detailed walls that rise 3500 m (12 000 ft) above its surface. This grand formation appears to perch atop its own highland region at the northern edge of Mare Nubium. Many sections of the walls are straight or scalloped, giving the crater's ramparts the outline of a polygon. Wilkins and Moore identified twelve "straight" sections in the ramparts. The northeast wall is broken by a deep notch that appears crater-like. A particularly deep scallop dents the east wall, and the crest of this scallop casts a very long shadow down the inner slopes of the east wall.

Copernicus A is a well-formed crater notched into one of the terraces, on the inner east wall, due west of the scallop feature. Surrounding the irregular walls is a more circular blanket of bright ejecta . "Only" 800 million years old, Copernicus is well preserved and young by the Moon's standards. All around the inner walls there is much evidence of huge landslides, where the debris dug out by the giant impact that created the crater has long since collapsed and subsided. In many places these landslips show up as arcs. The terraces merge with the crater floor at foothills that show signs of considerable erosion.

The interior of Copernicus, 65 km (40 miles) across, is more perfectly rounded than its walls. A double ridge of mountains running east–west with a wide valley between them divides the floor into northern and southern halves. The north floor of Copernicus is smoother than the southern half, the latter roughed out by numerous hills and ridges. There are numerous crater pits in both halves, but they are easier to make out in the north, where there is less competition from hillocks reflecting the glare of sunlight. Of the central mountain groups, the northernmost is the most complex in appearance – Goodacre says that this grouping consists of three major mountains "with several minor elevations associated with them." That's a fair description.

You should look for the peaks of these mountains ten days after new moon, when the first rays of sunlight catch them. You will be amazed at how the scene changes from one hour to the next. A day later and you'll be able to see the whole mountain masses in all their glory. These peaks are of modest height – the highest is around 750 m (2400 ft). The southernmost ridge is much longer than the grouping to the north that we just discussed; at its center is a mountain that at times looks like a perfect Egyptian pyramid. South of this ridge is another, lower ridge that curves toward the south rim; it may well constitute the remnants of a ruined crater ring formation.

The complex of ejecta from Copernicus is the most extensive of any walled plain on the Moon, and it consists of three different components. The first component, already mentioned, is the circular blanket of bright material that immediately surrounds the outer ramparts. This blanket is highly textured and slopes outward for at least 30 km (20 miles). Goodacre described the outer glacis of Copernicus as "broken by numerous radiating lava ridges," which betrays his belief that the feature was essentially volcanic in origin, but today we know this is not the case. The "lava ridges" are not that at all, but simply gouges in the Moon's surface dug out by rocky material violently hurled out by the impact that created the crater.

The second component consists of hundreds of secondary craterlets, many in short chains, that radiate from the main crater in all directions, but which are especially plentiful and spectacular to the northeast. Tracing these craterlet chains backwards, they all appear to converge on Copernicus, and there can be no doubt that they were formed by smaller chunks of rock that accompanied the principal impacting mass that dug out the main walled plain. Many of these craterlets are large, and obvious even in small telescopes and under conditions of average seeing (atmospheric steadiness). Many others will suddenly pop into view if you are privileged to have a larger telescope, or when the sky is particularly steady. I have counted several hundred that are visible in my 250 mm (10-inch) telescope. One of the most spectacular craterlet chains extends for 130 km (80 miles) to the northwest, all the way to the crater **Tobias Mayer C**; Goodacre mistakenly thought this chain was another example of the "lava ridges" that he associated with the outer slopes of Copernicus.

The third component of the ejecta is the magnificent ray system of Copernicus, best seen at full moon, and second only to the rays of Tycho. Goodacre was of the opinion that "any attempt at mapping them would be futile." Still, it's fun to try! These rays, which are superimposed on the older terrain lying beneath them, extend for hundreds of kilometers and are clear evidence that Copernicus is a relatively "fresh" lunar crater.

South of Copernicus lies the double crater **Fauth**, to which it appears to be connected by curved ridges that are probably the ruined walls of an intervening ancient ring. To the southeast are two large oval regions of dark mare material that are effectively enclosed by rays from Copernicus.

Tobias Mayer is a 35 km (22-mile) diameter walled plain lying at western edge of the Carpathian Mountains, accompanied on its eastern side by an almost perfectly round, deep crater known as **Tobias Mayer A**. The walls of Tobias Mayer are irregular, and there is a big hole in the north wall and a similar break through the south wall. You will easily be able to see a deep ravine in the inside northwest wall; more challenging is a rille just outside the west wall – it runs north–south, paralleling the wall, which itself appears strangely squared-off at this point. The floor of Tobias Mayer offers the observer some interesting though subtle features. There is a little group of hills just north of center, and more hills just inside the northeast rim. Just west of center is a weird group of four craterlets, the largest of which is very elliptical. East of center is a tiny gorge. Due south of the main crater appear several crater-cones. South-southwest is a large, elliptical ruined ring formation designated **Tobias Mayer P**, enclosed by scattered masses of mountains. Beyond this, the terrain smoothes out onto the Imbrium plain. Southwest is another such ring, next to the very circular but lava-flooded crater **Tobias Mayer B**. East of this formation, lying between it and the main crater, is a deep rille oriented in the same direction as the ravine in the northwest wall of the main crater.

Milichius, a small, bright 16 km (10-mile) wide crater due west of Copernicus, has walls that barely protrude above the surrounding mare plain. Look for the very low lunar dome due west of Milichius capped by a tiny crater pit; it is a most remarkable sight. South of Milichius you will encounter a number of mountain ridges that extend to Hortensius. The highest of these ridges is about 1000 m (3000 ft).

Mare Imbrium, the "Sea of Rains", is the largest of the Moon's lava-flooded plains. Lunar geologists believe that this great basin is the result of the Moon being struck about 3.85 billion years ago by an asteroid approximately 100 km (60 miles) in diameter. Since Imbrium measures 1200 km (750 miles) from east to west and 1100 km (680 miles) from north to south, you can see that an impacting body of relatively modest dimensions can carve out a hole many, many times its own size, thanks to the tremendous kinetic energy of the incoming asteroid. To fully understand the physics behind impact events, I would highly recommend a visit to Meteor Crater, located about an hour's drive east of Flagstaff, Arizona. In 1995 I had the privilege of hiking to the floor of the crater and making a detailed study of its structure. It was hard to believe that this 1.2 km (0.75-mile) structure was excavated by an impacting body no larger than a delivery truck.

Mare Imbrium is bordered on the north by the Alps, the Teneriffe Mountains, and the Jura Mountains, to the south by the Carpathian Mountains, and to the southeast by the Apennines. Its western shore merges with another lava plain, the Oceanus Procellarum. Three times the size of Mare Serenitatis and five times larger than Mare Crisium, the Imbrium plain contains many interesting objects, several of which are discussed in this chapter, among them Archimedes, Timocharis, Pico, and Piton. This chapter describes fewer objects than the chapters on the other three lunar quadrants, and there is a reason for this imbalance. The Mare Imbrium impact was so devastating that it wiped out many of the pre-existing features in the northwest quadrant of the Moon, features that are now forever lost to us.

Overall, as Goodacre pointed out, Mare Imbrium "nowhere shows any departure from a general level," though he did notice that its surface "is traversed in all directions by low curved ridges." Its color is a nearly uniform gray, but it is easy to see lighter areas here and there across its surface. South of Archimedes is one of these bright areas, located in an expansive hilly region that protrudes like a cape onto the southeast shore of Imbrium. The lava floods of this basin have left many partially submerged, or "ruined" crater formations whose original floors shall never be known. The area south of Plato is well known for having the largest display of wrinkle ridges to be found anywhere on the lunar surface.

The **Teneriffe Mountains**, a very bright range to the west of Plato, are situated on northern margin of the Mare Imbrium plain. At least one of the peaks in this range attains an altitude of 2500 m (8000 ft); it lies on the west rim of a large ghost crater immediately south of Plato.

Pico is easily one of the best examples of an isolated mountain mass on the lunar surface. This bright white peak may be found on the southern rim of a ruined crater ring, formerly designated "Newton," that lies south of Plato. It is about 2500 m (8000 ft) high and has been described by Walter Goodacre as "an immense obelisk." Its steep slopes are gouged with deep ravines, and tiny craterlets speckle the base of the mountain. The appearance of the mountain changes with the Sun angle, and Goodacre thought he was witnessing a "snow-storm" advancing across it, when all he was really seeing was the Sun playing tricks with the light on the hard faces of the formation. About 55 km (35 miles) south is another mass of mountains, but these are only half as high as Pico.

Mt Piton, similar in many respects to Pico and the isolated peaks of the Teneriffe Mountains, is a lonely bright mountain rising to over 2000 m (7000 ft) at the northeast edge of the Imbrium plain. It appears to be double, and its summit is capped by a craterlet. To the south lie a few craterlets and low hills, but the whole region is rather barren. Several historical lunar observers thought they saw changes in this feature that they variously ascribed to clouds or frosts, but in reality the appearance of almost all lunar mountains will change rapidly with the advancing terminator, and this effect is most dramatic for isolated peaks like Piton. Figure 6.1 is a splendid drawing of Mt Piton, showing many delicate details in the region. Note the very long and dramatic shadow cast by the mountain to the west.

Archimedes is the largest walled plain on Mare Imbrium, measuring 80 km (50 miles) across its interior. Its floor has been completely overrun by lava flows from Imbrium, so the height of its walls averages only a modest 1200 m (4000 ft) above the surrounding terrain, and its floor is featureless but for a few brightish streaks and tiny crater-pits on higher ground near the rim. The bright streaks are probably part of the rays from nearby the crater Autolycus. Lava has so filled the interior of Archimedes that its floor has now risen to a level only 200 m (650 ft) below the outside environs. A few peaks reaching as high as 2250 m (7400 ft) manage to rise above the fairly low rim. Like most formations of its type, the inner walls of Archimedes slope more steeply than the outer walls. The outer ramparts are about 11 km (7 miles) wide. A deep valley, described by some observers as a crater chain, rests atop the west-southwest wall and some-times makes the wall look doubled. The floor is so smooth that Wilkins and Moore aptly described it as "mirror-like," and indeed it often takes on an almost glassy appearance, especially at low Sun angles.

Despite the extensive lava floods, there is clear evidence of terracing all around the insides of the walls, and there are broad, gently sloping land-slips on the outer ramparts, splashed with bright splotches of ejecta , especially to the east. The walls have almost a braided appearance, like a length of rope. The coarse ejecta continues away from Archimedes across Palus Putredinis, which would

otherwise be smoother. As you look south and southeast towards the Apennines, you will be able to make out two or three rilles – more if seeing conditions are favorable. Some of these rilles appear to radiate from Archimedes, but are not necessarily part of its ejecta complex; rather, they are probably related to similar rilles that run parallel to the foothills of the Apennines. All of these rilles are likely just cracks in the Imbrium surface caused by the tremendous weight of the extensive lava flows that gradually filled in the basin over millions of years. The region south of Archimedes is rough and hilly, consisting of foothills that gradually become part of the Apennines. A deep crater with a barely discernible central peak, formerly known as **Gant** but now labeled **Archimedes A**, lies to the south-west. To the north is a very bright isolated mountain mass and a sinuous feature that could be either a wrinkle ridge or a rille partially filled with lava. Northwest are the **Spitzbergen Mountains**.

Timocharis is a freshly formed 40 km (25-mile) crater marooned on the Imbrium plane, west-south-west of Archimedes. The degree of terracing along the inner walls of this formation, especially on the western side, is nothing short of astounding for its overall size – there is hardly any land between the innermost landslips and the central craterlet, so that the terraces dominate the whole scene. The central craterlet is surrounded by some modest hills. This is probably the remains of a true central peak that was destroyed by the impacting body responsible for creating the central crater. The terraces give the impression that Timocharis is not very deep, but that is an optical illusion, for the floor lies a full 2100 m (7000 ft) below the crest of the rim. Although the terrain around the crater looks very dark at low Sun angles, as the Sun climbs higher a faint ray system appears around Timocharis, coincident with radial ridges that surround it.

La Hire is a 1500 m (5000 ft) high group of mountains northwest of Lambert; it is 19 km (12 miles) wide at its base, and its easternmost peak is capped by a craterlet. It is bright at both sunrise and sunset, but gets washed out at higher Sun angles. Goodacre hypothesized that La Hire and other isolated mountain masses of its kind "resisted erosive action better than many other formations on the Mare Imbrium." In truth, these isolated moun-tains survived the lava floods simply because they were lofty enough to protrude above the floods, while lower formations were submerged.

Lambert, a 29 km (18-mile) wide crater in the center of the Mare Imbrium,. has broad polygonal walls with generous terracing, a gap in the north rampart and a crater on the crest of the south rim.

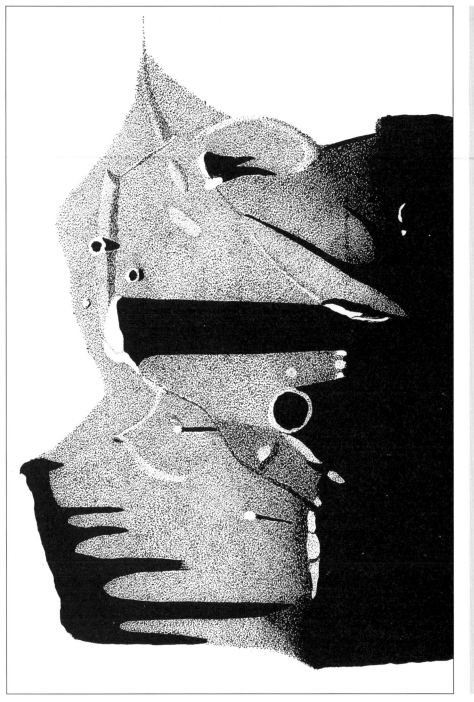

Figure 6.1.
Mt Piton, captured at a low Sun angle, casting a long shadow across the center. *Illustration by Colin Ebdon.*

The whole north half of the inside wall is dominated by a large landslip. In the center is a long ridge running southwest–northeast and capped by a little crater. This ridge is actually just part of a circular rim of hills that are punctuated by at least one other craterlet. Remnants of ejecta gouge the lunar surface outside Lambert in all directions, though these fea-

tures are subtle and require good seeing conditions to make out clearly.

Euler is a very beautiful medium-sized crater west of Lambert, and similar in many respects to its neighbor. It is 31 km (19 miles) across and 1800 m (6000 ft) deep. The floor has a number of objects worth looking at, including a central mountain dimpled by a crater, ridges formed into arcs that are part of the inner wall terracing, and some hills just inside the south rim. The western wall is very distorted and out-of-round. A fine rille dotted with craterlets runs southeast; to the southwest are several groups of mountains, hills, and isolated peaks. Many of these are rounded, and a few have crater-pits on their summits. The lunar surface to the north is very, very smooth, with only a very small number of inconspicuous craterlets marring the otherwise pristine lava beds.

Carlini is a small, distinct crater 8 km (5 miles) across and 600 m (2000 ft) deep in the middle of Mare Imbrium, south of Sinus Iridum. Many craterlets surround this little object, which also has a small hill at its center.

Helicon and Le Verrier are a nice pair of craters lying along the northern shore of Mare Imbrium. t first these two formations look a little boring, but under high magnification you can make out quite a bit of detail in each. **Helicon** is 21 km (13 miles) wide and 1500 m (5000 ft) deep, with a central bowl crater. Its floor, which appears to be lower on the west side, is accented by a series of arc-like ridges that are somewhat concentric with one another. A crater discovered by Patrick Moore in 1951 using a 215 mm (8½-inch) reflector dimples the southwest wall. This shows that you do not have to use huge telescopes to make interesting discoveries about the Moon, which offers more detail than any one observer could hope to fully appreciate in one lifetime. The careful observer will detect terraces along the inner walls; the terraces on the east side are very weird and irregular-looking. The crater is set off from the surrounding dark mare by large splashes of bright material.

Le Verrier is the companion crater to Helicon, lying about 30 km (20 miles) to the east. It is 18 km (11 miles) across and about 1400 m (4500 ft) deep. Its walls rise only 450 m (1500 ft) above the Imbrium plain, however, so it is a fairly low-contrast object. Its floor is dark and devoid of any major features. A few scattered hills are visible in larger amateur instruments. Some terracing is visible along the inner walls, but it can be difficult to see even this aspect of the crater.

Plato is a large (100 km, 60 miles), extremely dark lava-flooded walled plain located at the northern margin of Mare Imbrium, between the Alps and the Teneriffe Mountains. Plato is one of the major guideposts on the lunar surface. Its floor is at first glance completely smooth and featureless, but at high power and under conditions of steady seeing you should be able to see at least four minute craterlets dimpling its interior. Interestingly, these lie in

an arc along the crater's north–south axis, so they are fairly fresh, having come along after the lava floods. These floods were very extensive, as few if any features survive even at the margins of the crater's interior. Lunar transient phenomena (LTPs) have been reported in Plato, usually said to take the form of dust clouds that obscure one or more of the interior craterlets. Most of these LTPs are probably illusory – the craterlets of Plato are already at the threshold of visibility in many amateur-sized telescopes, so minute variations in seeing can cause the craterlets, or some of them, to appear to blink in and out of existence before your very eyes! I have witnessed this phenomenon several times, yet not once have I ascribed the effect to "obscuring clouds," and despite thousands of hours spent observing the Moon since 1970 with almost every size of instrument imaginable, I have never seen such a cloud or anything else I could clearly ascribe to a gaseous emission from the lunar surface. That is not to say such events do not occur, but I would suspect that many "obscurations" of lunar detail may be explained more properly by fluctuations in sky conditions rather than by geologic events on the Moon itself.

For many years, classic observers of the Moon propagated the belief that the floor of Plato actually *darkened* as the Sun rose higher over the formation, which is of course contrary to what we know about the behavior of reflected light. Since light is always reflected away from a surface at the same angle at which it strikes that surface (the angle of incidence), unless the floor of Plato were violently tipped to one side it should brighten (reflect more light) as the Sun rises above it, since more of the incident light would find its way back toward the Earth-bound observer. Careful studies have shown that this is indeed the case – the floor of Plato is fairly level, and not convex, and in fact the floor does brighten as the Sun rises higher, like virtually every other lunar crater. The effect noticed by the classic observers is just an interesting optical illusion: a contrast with the walls, which brighten so dramatically as the Sun angle increases that the floor appears darker by comparison. You can duplicate these experiments for yourself to understand why the classic lunar observers thought as they did.

The walls of Plato are richly braided, especially on their western sides. They rise 900–1150 m (3000–3500 ft) above the smooth floor, but their overall altitudes above the surrounding lunar terrain can be in the neighborhood of 1500 m (5000 ft) or more. The eastern rim crests at a maximum height of over 2530 m (8300 ft), so clearly the lava inside Plato is actually on a higher plane than the lava in the mare outside the crater. This and two other peaks on the east rim, that rise to heights of 2225 and 1580 m (7300 and 5200 ft), respectively, cast long shadows across the crater's floor at sunrise. A very bright peak on the west rim, at the apex of the famous triangular formation known as ζ, towers 2230 m (7500 ft) above the rest of the formation. This triangular area is actually part of the aforementioned braided walls.

A deep rille runs westward from this region, down the outer slopes to a craterlet named **Plato P**. The outer walls are really just huge piles of rubble tumbling down to the rolling foothills of the Alps. Also outside the west wall is a series of four craters, beginning with **Plato A** that lie in a row heading northwest. There are actually six smaller craters here, too, that branch out from Plato A, giving the whole assemblage the form of a human stick figure, with Plato A as the head, the six small craters serving as the arms, and the rest of the craters making up the torso. The whole region is strewn with small craters, many partially overlapping one another, and bright little mountains that are outliers of the Alps range.

Figure 6.2 shows two realistic drawings of Plato, one at sunrise, the other at sunset, so that you can see the shadows cast by the walls on each side of the walled plain. These drawings show a very modest amount of detail on the crater's floor, apart from the shadows, in contrast to depictions of the formation by the many classic lunar observers who were apparently compelled to outdo one another in discerning faint features inside Plato's walls. Much of the detail they depicted was, not surprisingly, simply not real. For example, William H. Pickering found seventy light-colored spots inside the form-

ation, but there is simply no evidence, photographic or otherwise, that anything of the kind can be seen. Depicting spurious detail is to be avoided at all costs, for it creates yet another problem: because many of the older drawings of Plato looked so different from one another, many observers assumed that the crater was undergoing "changes."

Harpalus, a 40 km (25-mile) wide crater at the western margin of Mare Frigoris, has lofty 5000 m (16 000 ft) high walls that make it one of the very deepest on the Moon. Harpalus has not one central peak, but four. The whole floor is rough-hewn with impact debris, and the inner walls are strongly terraced, with great landslips that cascade down to the crater floor. Three craters of nearly identical size lie next to one another atop the south rim, an old ruined ring lies just outside the southeast wall, and two tiny craterlets are visible on the crater floor – one just inside the north rim, the other, somewhat larger, due south of the central mountain group. Look for a low ridge concentric with the northeast inner wall – Patrick Moore discovered this feature in 1952. Fifteen years later, one of the Lunar Orbiter spacecraft confirmed its existence with a fine photograph.

Sinus Iridum, the Bay of Rainbows, is a 240 000 km^2 (92 000 square mile) dark splotch of mare material that bites into the north shore of Mare Imbrium. This, the most magnificent of all the lunar "bays," is probably a discrete impact site, separate from Imbrium, as suggested both by its semicircular northern border formed by the Jura Mountains, and by the wrinkle ridges that show how lava from Imbrium overran the bay's southern border. The logical conclusion is that Iridum

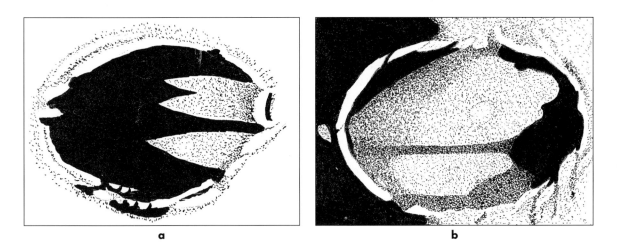

a   b

Figure 6.2.   **a** Sunrise and **b** sunset on Plato. *Illustration by Sally Beaumont.*

resulted from an impact that happened after the catastrophic impact responsible for creating the Mare Imbrium but before the Imbrium lava floods. It would have been interesting to watch these lava flows, which took place over a period of millions of years, gradually break down the southern wall of Iridum.

The floor is smooth, as you would expect, with just three inconspicuous craterlets, the largest designated **Iridum A**. The **Jura Mountains** are of modest height on their northeast and southwest rims, but much higher in the middle, with at least one peak towering 6000 m (20 000 ft) over the bay's interior, near the crater Sharp. A second peak nearby to the west is almost as lofty. It is not hard to detect the many huge piles of rubble inside the rims of these mountains, and on the outer slopes are numerous valleys that cut through to the foothills below. A 40 km (25-mile) crater, **Bianchini**, crests the north rim of the Juras. Bianchini is deep, its south wall rising to 2250 m (7400 ft); its west wall is even higher – 2600 m (8500 ft) above the floor. Its interior presents considerable detail to the lunar observer, including a central mountain, two ridges that are probably landslides from the rims, and craterlets.

Sharp is a deep, 35 km (22-mile) wide crater on the northwest slope of the Jura Mountains. There is much to see inside this crater, including several hummocky areas and a small central mountain, a ridge from the east, and terracing on the inside west wall. North of Sharp runs a short crater chain that terminates at the associated crater **Sharp A**.

The **Heraclides** and **Laplace Promontories** are two capes which mark the entrance to the Bay of Rainbows, Sinus Iridum. Laplace, rising more than 2750 m (9000 ft) above the bay, guards its northeast mouth; Heraclides guards the southwest entrance. Laplace is a plateau overlooking massive cliffs that mark Iridum's edge; its summit is pocked by no fewer than four craterlets. Look for several strong ridges that head southeast from the southernmost end of Laplace. Heraclides is a triangular group of mountains 1200 m (4000 ft) above the bay's floor. Its eerie resemblance to a woman's head caused the classical observer Cassini to dub it the "Moon Maiden." This resemblance is not apparent under all light angles; like most isolated peaks, the appearance of Heraclides changes rapidly with the rising and setting Sun. Your best chance of catching this phenomenon is just after lunar sunrise.

Condamine is a rhomboidal walled plain, about 50 km (30 miles) wide with 1200 m (4000 ft) high walls. There is a pass through the north wall, and the walls on the west side are a little higher than else-

where around the rim. On the outer northeast slope, three sizable depressions are evident, and a couple of craterlets on the crest of the rim above. The floor is equally interesting, with a high ridge to the north, and a low, rounded hillock to the south; to the southwest are four "spurs" attached to the wall, and in the northwest corner appear three strange squarish enclosures. A number of valleys surround this walled plain, meandering down to the shores of Sinus Iridum.

John Herschel is technically a walled plain, though the walls are barely noticeable in places; this 145 km (90-mile) wide crater is found in the northern hinterlands of the Moon's northwest quadrant, north even of Mare Frigoris. The classic English selenographer Edmund Neison characterized this area as "no true formation, merely a portion of the surface surrounded by elevated regions." John Herschel has a tall mountain on its eastern border, and its "walls" enclose many craters, the largest of which is north of center. Because of foreshortening induced by the formation's closeness to the north pole, this crater, designated **John Herschel C**, appears nearer the center than it actually is. For the same reason, two large overlapping craters on the southern border appear very elliptical, though in fact they are quite round.

A fine rille wanders along the inside of the east rim; it throws off two branches that head in a westerly direction towards John Herschel C. But perhaps the most remarkable objects on the floor of this formation are a series of curved gouges in the surface that run north–south just west of center. A surprising number of craters dot the whole southern highlands. Southwest of the principal enclosure is the crater **Robinson**, distinguished by an odd valley that protrudes from its northeast rim. West of John Herschel rises a unique, snaking high ridge that is punctuated in many places by tiny craterlets. This is actually the shared rim of the walled plains known as **Babbage** and **South**, structures that appear very similar to John Herschel in many respects.

Encke and Kepler are a most striking pair of strange polygonal craters perched atop elevated terrain west of Copernicus. **Encke** is 32 km (20 miles) wide with comparatively low walls that rise a mere 600 m (2000 ft) above the surrounding surface. There are several unusual features of this walled plain, not the least of which is its hexagonal outer wall, which appears very thin and sharply defined all around. On closer inspection, however, it becomes obvious that the inner slopes of this rim are heavily terraced, the terraces looking more like a series of disconnected ridges that intrude far onto the interior floor. It is difficult to imagine how some of them got to be where they are! The classic guidebooks all mention a "central ridge running north and south," but I find no evidence of this. Rather, I see many small hillocks in a generally rough central area, but no true central mountain mass or ridge. The ridges all lie closer to the inner rims – three very lofty ridges are just inside the east wall, separated from lower ridges nearer the center by a broad valley, and another prominent group of ridges may be found inside the southwest rim.

A remarkable forked rille runs from north to south, but well west of the middle of the crater. It is interrupted by a craterlet in the northwest corner of the floor. Dark craterlets also indent the west and south rims, and a very curious funnel-shaped valley lurks just beyond the southwest wall. Immediately south of Encke is an isolated group of rounded mountains, and farther to the south you will see a much larger and complex mountainous terrain that forms part of a ruined ring formation, several times the size of Encke itself. Encke appears to lie just inside the northeastern corner of this old ring, whose interior is an interesting study of dark solidified lava splashed with bright ejecta, much of it probably from Encke. Due east of Encke you should easily spot a row of three small, deep craters all atop one another, their rims overlapping. The terrain north of Encke is extremely rough, giving a very pebbly appearance. The "pebbles" are innumerable small hills.

Kepler is also 32 km (20 miles) across and, like its neighbor Encke to the south, it has somewhat squared-off walls with razor-sharp rims rising to heights of 3000 m (10 000 ft) in places, and massive landslips spilling far onto its floor. Kepler is located at the north end of the same highland region that host Encke, like an island floating on the Oceanus Procellarum. If you look at Kepler's walls at high magnification, you will notice, however, that its walls are really not as straight in sections as Encke's; rather, they are warped and offset in places, especially at the northeast and northwest corners. A deep valley completely disrupts the northwest wall.

You will also want to examine the floor of Kepler at high power, where you will notice three mountain masses. Two of these lie just east and west, respectively, of the crater's center, separated by a very bright valley. The third is just inside the northern rim; it spills southward, and joins the westerly "central" mountain by a low, bridge-like structure. Immediately south of the two central mountain masses there appears to be a perfectly circular crater ring punctuated by a tiny central peak. To the north of Kepler radiate three separate mountain ranges, looking rather like fingers. They are just part of

Kepler's massive ejecta blanket, which brightens the surrounding mare material out to 50 km (30 miles) or so in all directions, but especially to the northeast. Outside the ejecta blanket is Kepler's famous ray system, which extends across the lunar surface in an immense starburst pattern for 150 km (100 miles) or more. Thanks to this amazing ray system, Kepler is surprisingly prominent and easy to locate for such a modestly sized crater. Figure 6.3 shows not only Encke and Kepler, but also the terrain between them.

Reiner is a very interesting and weird walled plain that contains some unique formations on its floor, but is best known for the even more puzzling feature to its west known as Reiner γ. Reiner itself has bright,

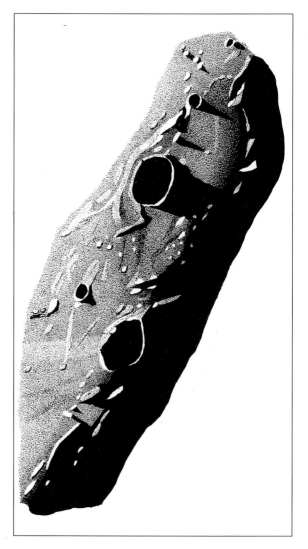

Figure 6.3. Encke (above center) and Kepler, and the surrounding area. *Illustration by Colin Ebdon.*

terraced walls that soar 3000 m (10 000 ft) above a dark and tortured floor. The floor is chaotic. A bright, double-peaked central mountain mass looks normal enough, but two strange, curving gorges strike south to a dark, smooth "oasis" that is bordered to the east by what look like a series of small wrinkle ridges and to the west by what are most likely massive landslips. Even these landslips do not look quite normal, however, having a very wrinkled appearance themselves, much like an elephant's hide. They are very bright. Another sizable gorge cuts into the southern floor.

Reiner displays an impressive blanket of bright ejecta for a crater of its size (a mere 32 km (20 miles). From the south rim runs a wrinkle ridge, which Wilkins called a cleft, terminating at the north "wall" of an almost completely submerged crater ring. This ruined formation, known as **Reiner R**, is fully twice the size of Reiner, and looks more like a pair of parentheses, only the curves of the east and west walls having survived the lava floods of Oceanus Procellarum. There are many wrinkle ridges in the vicinity, especially between Reiner and the crater **Hermann** some distance south.

Despite this wealth of interesting detail, Reiner itself is often overlooked by lunar observers, because of their fascination with the unique splash of bright material known as **Reiner γ**, so called because of its superficial resemblance to the Greek letter gamma. It almost certainly consists of the same highly-reflective material as the lunar rays, but unlike ordinary rays, which are straight, this flat feature is anything but straight. One hypothesis for the formation of Reiner γ is the ejection of impact debris into low lunar orbit, where it is able to travel around the lunar globe in opposite directions until is falls back to the surface. It is conceivable that some of this ejected material collides with similar material circling the Moon in the opposite direction, making weird shapes as it falls to the surface below. At least one other example of such a feature is known – in Mare Marginis, on the Moon's extreme eastern limb, which lunar geologists believe was formed in the way just described after the impact that created the Mare Orientale basin on the exact opposite side of the Moon.

Marius, an ordinary crater 42 km (26 miles) across with 1200 m (4000 ft) high rims, would be of no special interest to lunar observers except that directly to its west lie a vast field of volcanic domes, unique on the lunar surface. Amateur lunar observers should have no trouble spotting dozens of these shield volcanoes, which look just like their Earthly counterparts. Contrast the appearance of these domes to the ordinary central mountains of

lunar craters, which early observers also thought to be volcanoes. Ironically, many of these observers did not recognize the volcanic nature of the Marius dome field, dismissing these objects as "low, convex hills." But the Marius domes really do look like proper volcanoes, and they even have volcanic vents, in the form of crater pits on their summits. I have detected many of these volcanic vents in my 250 mm (10-inch) telescope, for which excellent seeing is required, but they are easier to find in larger apertures. Two planned Apollo missions (Apollos 18 and 19) were supposed to explore this fascinating region in depth, but they were sadly canceled, preserving indefinitely the mysteries locked in these volcanic rocks.

Aristarchus, together with its neighbor Herodotus and the collapsed lava tube known as Schröter's Valley, forms perhaps the most studied group of features on the Moon's surface. Aristarchus is rightfully famous for two reasons – it is easily one of the youngest of all the lunar craters, and, thanks to its youth, its freshly exploded rock makes it the brightest object on the Moon. The famous German-English astronomer Sir William Herschel believed it to be an active volcano, but we now realize that it is really an impact site, like all of the other lunar craters that look like it. Herschel would be pleased to know, however, that the region of Aristarchus does indeed show much evidence of ancient volcanic activity, including the lava tube known as Schröter's Valley and numerous volcanic domes on a plateau north of Aristarchus referred to as the Aristarchus Uplift.

Aristarchus is 47 km (29 miles) across and, according to Wilkins and Moore, its walls have "a decidedly polygonal contour, especially on the [west] – though the departure from a regular curve is much exaggerated in the drawings of many observers." The walls are not particularly high, at only 600 m (2000 ft) above the surrounding lava, but they are sharp and uneroded, so they show the formation in high relief. Very detailed terraces are beautifully preserved along the inner walls, though the powerful glare inside the crater can make these details, as well as the diminutive central peak, difficult to see. The walls are quite complex, and Elger mentioned the existence of many "spurs and buttresses," especially on the south walls. You should also look carefully for the famous dark vertical bands that appear on the inner walls. These radial bands, most easily spotted inside the western wall, are probably the result of offsetting from the landslips that created the terraces, too. Wilkins and Moore identified nine of them. I strongly recommend using a variable polarizing filter (see Chapter 8) when viewing Aristarchus

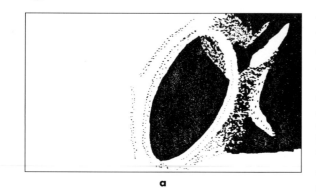

a

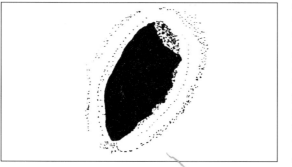

b

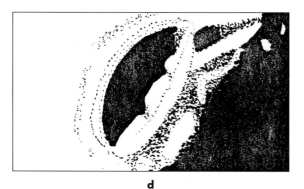

c

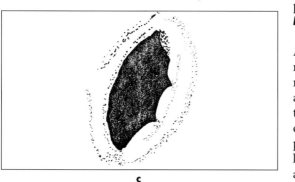

d

Figure 6.4. The appearance of bands in Aristarchus as the Sun angle increases. The drawings were made during a two hour period when the co-longitude was **a** 48°.6, **b** 49°.1, **c** 49°.4, and **d** 49°.5. *Illustrations by Sally Beaumont.*

to cut down the glare and disclose these features to their best advantage. Figure 6.4 depicts the emergence of the dark bands from shadow inside Aristarchus.

Aristarchus is the center of a fairly bright ray system – the most prominent system of rays run south and southeast – and its outer slopes are dusted with a generous blanket of bright ejecta. But both of these features pale in comparison with the blindingly white crater rim itself, so bright that you can see Aristarchus without difficulty by the "earth-shine" a few days after new moon. I know of few other lunar features that can be detected so easily when they are beyond the sunrise or sunset termina-tors. The region north of Aristarchus is called the Aristarchus Uplift, and is worthy of careful study. Besides Schröter's Valley, which is discussed below, this plateau is geologically very diverse. There are many delicate rilles radiating southward, but also numerous wrinkle ridges crossing the lava in various tortured shapes. There are a few crater rings that are filled with solidified lava to different degrees, like the flooded ring **Prinz** northeast of Aristarchus. This ring, whose southern wall is com-pletely overrun, lies at the foot of the **Harbinger Mountains**.

Herodotus is the companion crater to Aristarchus, at 37 km (23 miles) nearly as large, but much older and less well preserved than its bright neighbor. Its 1200 m (4000 ft) high walls are eroded, and its interior is mostly filled with solidified lava that has blanketed many of the details we would otherwise expect to see here. There is no central peak, at least none high enough to have survived the lava flooding, and the walls also appear devoid of any major features, though there was probably the typical terracing before the lava came. Two rocky spurs jut from the south wall; upon closer examin-ation you will realize these are the remaining walls of a crater whose south rim has been destroyed by lava flows. This crater appears to be resting on the south slopes of Herodotus, tilted down at an odd angle towards the plain below. A tiny craterlet, **Herodotus N**, is tucked inside the north rim. The eastern wall is very straight; this and the foreshort-ening effect give the crater a polygonal appearance. There is a gap in the north wall that leads to Schröter's Valley, the name of which records it dis-covery by Johann Schröter on October 7, 1787.

The terrain between Schröter's Valley and Herodotus is very hilly, and one group of hills, west of the main crater, appears to be the enclosure of an old crater whose west side is now gone. Further north you will you find a very prominent wrinkle ridge that runs southwest–northeast. To the south lie a number of ruined rings and ghost craters, some visible only because they have been highlighted by ejecta from Aristarchus, much as a pencil rubbing discloses the details of its underlying subject.

Schröter's Valley. Perhaps no object on the Moon's surface has captured the imagination of lunar observers and mappers like the lava tube known as Schröter's Valley, the best and most easily observed example of a lunar sinuous rille. Flowing from an oval depression (possibly a filled-in volcanic vent) known as the Cobra's Head just north of Herodotus, the rille curves for more than one hundred miles across the Aristarchus Uplift. If you peruse the classic lunar guidebooks, you will see that there was once a long-standing debate about whether the Cobra's Head actually opened into Herodotus or not – this question is probably academic, as it is a discrete structure in its own right associated with the lava flows that formed the valley, though it may have played a role in filling up poor flooded Herodotus.

After spilling out of the Cobra's Head, Schröter's Valley winds northward for about 50 km (30 miles) through a high plateau, then curves northwest onto a much smoother, lava-dominated terrain for about the same distance. It is here that the rille makes a weird zigzag, heading first southwest for about 6 km (4 miles), then abruptly turning back northwest again for 15 km (10 miles) or so, finally swerving southwest a second time, gradually tapering over a distance of some 80 km (50 miles) to a group of mountains northwest of Herodotus, where it is swallowed up by the rough terrain. On nights of excellent seeing I have seen a hint that Schröter's Valley does not actually end at the base of these highlands, but continues through them as a dark streak. Lunar Orbiter photographs confirm this to be the fact, though the rille is very delicate indeed.

Hevelius, also called "Hevel" in the older guidebooks, is a remarkable ringed plain on the Moon's equator, visible at the extreme western limb. Some 110 km (70 miles) in diameter, its floor shows an incredible amount of detail that almost defies description. If you want to see examples of almost every type of basic lunar formation, look no further than the floor of Hevelius, which is crisscrossed by long, deep rilles, punctuated by mountains, and indented by many craters large and small. Because of its proximity to the Moon's western limb, foreshortening greatly affects everything we see within this feature.

You will have no trouble finding plenty of craters on the rims of Hevelius and on its floor. A large, shallow crater intrudes onto the northeast wall, with smaller craterlets inside it – see if you can spot any of them. A deeper, darker crater perches atop the southeast rim, and a huge crater interrupts the southwest rampart. A crater almost as large dominates the northwest corner of the floor of Hevelius, while smaller craters speckle the central, southern, and eastern parts of the floor. A triangular mountain mass is situated a little northwest of center, its peak topped by an elongated crater. The appearance of this mountain changes almost hourly.

The system of rilles on the floor of Hevelius is quite complex. Two long, fine rilles run from the north rim to the inside of the southeast wall, while another cuts north–south just east of center, and yet another zigzags from the southwest floor to a craterlet near the northeast wall. A much shorter, rille curves across the west floor, while another lies between the central mountain mass and the large crater on the northwest floor. At high power using large instruments, still finer rilles pop into view during moments of excellent seeing.

The eastern side of Hevelius is bounded by dark, menacing walls that rise to heights of 2000 m (6000 ft) or so; the west walls are significantly lower and more broken, but very bright. The north and south walls are very much eroded, almost flattened. Everywhere the walls are very broad and show complicated structure, with braided, rope-like rims and numerous depressions, valleys, and gorges. The inner walls show some evidence of the well-known "terracing" effect, but mostly consist of disconnected ridges, slumped mountains, and huge outcroppings of rocky debris.

The outer walls show more of the same, and the great ring of ejected debris that surrounds the crater is very asymmetrical, being much broader at the northeast and southeast corners than anywhere else, but difficult to detect on the west side where it merges with the adjacent heavily cratered highlands that separate Hevelius from **Hedin**. A huge rille emanates from the southwest wall and heads toward an interesting grouping of craters associated with the crater **Lohrmann**, which is very close to the southern rim of Hevelius. The much larger and wonderfully detailed walled plain **Cavalerius** is tangential to the north rim of Hevelius. Cavalerius shows all the features typical of any good lunar crater – thick, terraced walls, many mountains and hills on its floor, including a group of central peaks, and gaps in both the north and south walls.

Anaxagoras, near the north pole, is a fresh, young 51 km (32-mile) wide crater that, like others of its kind, is the center of a bright ray system best seen at or near full moon. These rays extend south to Plato. Like other recent craters, it has steep, finely-terraced walls that tower 3000 m (10 000 ft) above its rubble-strewn floor. The highlight of the floor is a 300 m (1000 ft) high central mountain that, despite its modest altitude, is very effective at reflecting sunlight and hence appears very bright. It is just one peak in a whole range that runs from the southwest rim to the base of the northeast wall. The southwest side of this range is actually forked. The terracing is widest on the northwest wall, which also has a spur protruding from its outer crest. The southwest rim is greatly disturbed by the mountain ridge that runs across the crater floor – it looks as though this range penetrates the rim at this point and continues southwestward some distance outside the crater. Anaxagoras overlaps the heavily broken-down, 80 km (50-mile) wide walled plain known as **Goldschmidt**, which lies due east. Due west of Anaxagoras is a parallel series of very rugged mountain ranges, all of them oriented north–south.

Reinhold is a 48 km (30-mile) wide walled plain south of Copernicus. Almost all of the 2750 m (9000 ft) high eastern wall has separated into two concentric ridges, leaving a deep ravine between them. Detached, broken ridges make up the inner west wall. Small hills are scattered over the crater's floor, but there is no true central mountain. Higher, rounded mountains abound outside Reinhold to the east and southeast. To the west and southwest the terrain is much smoother – lava plains streaked with bright rays from Copernicus.

Hortensius, a deep, well-formed, nearly perfectly round bowl crater northwest of Reinhold, is only 16 km (10 miles) across. The sharp-rimmed crater itself is unremarkable, but there are many volcanic domes in its vicinity, including a most interesting chain of them directly northeast of it. This group of six domes is arranged in three matching pairs, all in a row! I have seen crater pits atop five of the six domes, strong evidence of their volcanic nature. **Hortensius E** is a submerged crater just as large as Hortensius itself, but almost completely filled with solidified lava, lying between the main crater and Reinhold. Figure 6.5 depicts Hortensius and some of its nearby domes.

Timaeus is an odd-shaped crater, 35 km (22 miles) wide, marooned on the northern edge of Mare Frigoris. Its walls are extremely bright, and it has a well-defined central mountain ridge. The west wall is fairly round, but the southeast and northeast walls are straight segments that come together at a sharp point. Just east of this apex is an impressive mass of mountains, and still farther east you will find a series of long, rugged ridges, all running southwest–northeast. Timaeus has a modest ray system of its own.

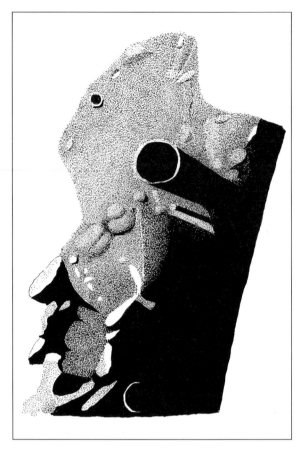

Figure 6.5. Hortensius and some of its nearby domes. *Illustration by Colin Ebdon.*

Fontenelle is an interesting 37 km (23-mile) wide walled plain on the northern shore of Mare Frigoris. Its walls are not that high, so the whole crater is really just a depression in the Frigoris lava plain. The walls are certainly not lacking in character, though, with a wide gorge distinguishing the west side, undulations on the east side, and a sharp craterlet on the crest of the southeast rim. Even more interesting is a 1000 m (3000 ft) high mountainous plateau that rests against the exterior of the southwest wall, topped by at least five sizable craterlets, two of which overlap, the other three in a group with rims just touching one another. An even stranger feature juts from the northeast wall to a craterlet about 6.5 km (4 miles) away – it looks alternately like a ridge or a pair of sinuous rilles. What do you think this feature is? The floor of Fontenelle is no less distinctive – in the center are two craters, one an oval hollow, the other a very round walled plain in miniature, with sharp but very low rims and even a very small central peak. Part of this crater's walls are obliterated, however. Several ridges cross the floor inside the rims, especially to the west, roughly (but not perfectly) concentric with the walls. Pay careful attention to a remarkable system of delicate rilles on the east floor, arranged in a pattern something like the letter C.

Pythagoras, one of the largest walled plains at 120 km (75 miles) across, is challenging to observe and very foreshortened by its proximity to the Moon's northwest limb. The view is nothing less than dramatic. The incredibly massive walls of Pythagoras tower 5000 m (16 000 ft) above its floor; they are varied and complex. The south wall consists of two great straight ridges running roughly south-west–northeast, coming together to make a V. Between the ridges is a very high, bright mountain peak. The west wall is extremely broad and gently terraced; many of the terraces are rather eroded. The outer rim of the west rampart is polygonal and contains two very long canyons where inner parts of the walls have separated and slumped. The north wall is ill-defined, an extremely inhospitable-looking assemblage of rugged mountains cut by steep valleys that drop precipitously to the floor below. The east rampart is characterized by high serpentine ridges with sharp crests and unimaginably deep gorges.

The floor of Pythagoras is distinguished by a group of central mountains that you can see a day or two before full moon, from the perspective of peering over the great crater's rim. This is a benefit of the foreshortening effect we have mentioned so often. Goodacre found "at least six or seven peaks" in this group of mountains, the highest of which reaches about 1500 m (5000 ft). Abutting the central mountains is a large crater which is often in deep shadow. It is curious that the north half of the floor is fairly smooth, in contrast to the south half, which is very rough, accented by many parallel ridges and lower hills.

The country around Pythagoras is also worthy of careful study. The southeast wall adjoins the very large, low-ringed formation known as **Babbage** at a very wide, very high mountainous region. You will see uncountable numbers of rugged ridges to the south and southwest of Pythagoras, while the terrain to the north and northeast contains more in the way of isolated, discrete mountains. Many of these are quite massive. Just beyond these mountains you will find two huge and ill-defined formations with low walls – **Anaximander** and **Anaximander B**. Just north of this pair of walled plains, and touching each of them, is the better-defined crater **Carpenter**, which features twin central mountain peaks.

Chapter 7

Lunar Features – Southwest Quadrant

Tycho, at the center of the Moon's brightest and most extensive system of rays, is the most obvious crater visible on the surface of the full moon, and therefore as good a place as any to start exploring the southwest quadrant. Tycho is a huge walled plain, 90 km (56 miles) in diameter, its walls soaring an impressive 3500 m (12 000 ft) above its floor, capped with peaks 1500 m (5000 ft) high. Wilkins and Moore describe the walls as consisting of a series of straight segments, and indeed this is how they appear, though overall the crater outline is pretty round. The segments were likely produced by differential slumping of the debris blown free from the crater's center at impact after it piled up to make the inner walls of the crater. Landslips occurred at different times and places around the crater walls, and not in a perfectly smooth circular pattern.

Take your time to carefully explore the structure of the inside walls of Tycho, which are amazingly complex. You will of course want to do this at low Sun angles. When the Sun is anywhere close to being overhead of Tycho, the crater and its rays shine with such a blinding light that almost all detail is lost. The terracing is very extensive, covering almost half of the floor, and very irregular – instead of perfect concentric circles, the landslips and ledges inside the crater are very much broken up into a collection of wriggles, arcs, and short straight segments that are almost nowhere continuous to any great degree. Look for variations in altitude here, also the contrast between some ridges which are rounded, and others which are sharp and uneroded-looking.

Tycho is noted for its twin central mountain peaks. They are not actually perfect twins, as one is bigger and, at 1500 m (5000 ft), taller than the other, and the smaller peak is capped by a prominent craterlet. There are really no other mountains,

craterlets or rilles worth mentioning on Tycho's floor, just scattered rubble piles. The terrain around the outside of Tycho is a different story, however, with innumerable craters of all sizes, many of them in chains, pocking the landscape. One particularly interesting crater chain southwest of the main crater is made up of at least seven different craters with overlapping rims, all in a row. The craters at each end of this formation are by far the largest, giving the chain the appearance of a dumbbell. The many craters surrounding Tycho have two characteristics in common – most are rather shallow, though they are not overrun with lava and do not have broken or disturbed walls, and most lack central peaks. Figure 7.1 is an instructive series of four drawings demonstrating the changing face of Tycho as the Sun rises higher over it.

As we have mentioned, the ray system of Tycho makes this walled plain famous. Wilkins and Moore found over one hundred rays emanating from the formation. It is no wonder, then, that at full moon the rays from Tycho overwhelm many nearby form-ations, including the walled plain Maginus, causing the nineteenth-century German selenographer Johann Mädler (1794–1874) to exclaim "the full moon knows no Maginus!" The most prominent of Tycho's bright rays streaks northeasterly across Mare Serenitatis, where it contrasts with the very dark lavas. You can't miss this ray, which intersects the crater **Bessel**, or the long rays that stretch all the way to Bullialdus (northwest) and Fracastorius (far east). Another bright ray extends to the southwest toward Scheiner.

As with many ray systems, the rays of Tycho do not continue up to the very rim of their parent crater, but they are separated from it by a dark "collar" of basalts. Inside this dark ring is a bright ejecta blanket, then the crater's rims themselves.

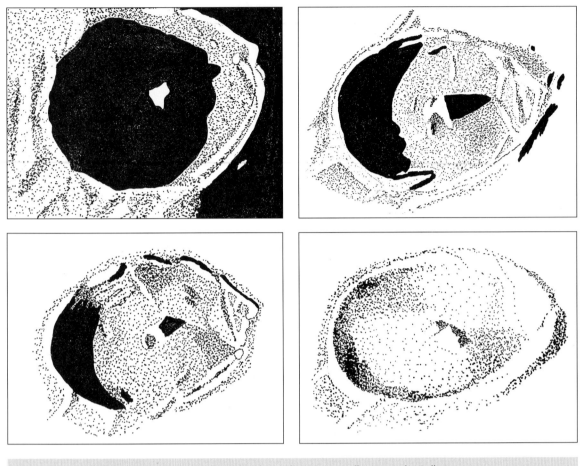

Figure 7.1. Four drawings of Tycho as the Sun rises. *Illustrations by Sally Beaumont.*

There are at least three different theories to explain the dark collar of Tycho and other craters like it. The first theory is that the collar is just a contrast effect, a region that is darker than the large amounts of bright ejecta immediately surrounding the crater's rim and the bright highland terrain that lies at a greater radius from the crater's rim. The second theory is that Tycho is situated atop another, older impact site, the dark collar giving away this older impact basin's location. A third theory is that the Tycho impact itself brought the dark material to the surface from deep within the lunar crust. At any rate, the dark collar of Tycho illustrates just how much we have yet to learn about the Moon and its geologic history.

Like all rays, the rays of Tycho are mere surface markings, for they cast no shadows even at very low Sun angles, and so have no appreciable height whatsoever – they are just scourings of the surface by debris ejected from the central crater at the impact of whatever body – asteroid or comet – made that crater in the first place. Another interesting characteristic of rays systems, including Tycho's, is that the rays cannot be traced back to a single point inside the parent crater: instead, that the rays appear to radiate from several different points inside or nearby the parent crater, sometimes tangential to it. Why this should be is a mystery.

Finally, see how far you can trace each ray – you will find that they cross all sorts of landforms with equal ease, showing that they consist of material that was deposited on the surface and are not the result of some internal geologic process that brought them to the surface. Although their intensity may vary along their individual lengths, the rays are usually fairly continuous along those lengths; notice also that some rays are much longer than others. The ray system of Tycho, as well as the fact that its features are very sharp and well defined, show that it is a comparatively young, freshly formed crater. This hypothesis may have been confirmed by samples brought back by Apollo 17. Although the Apollo 17 samples were from many hundreds of kilometers away, there is evidence that some samples were of ejecta from Tycho, and, if true, the crater would be just over 100 million years old – "young" by lunar standards. Tycho stands in contrast, then, to the region of the Moon in which it is located – the so-called southern highlands – which is undoubtedly one of the most ancient regions of the lunar surface.

Pitatus, a large (80 km, 50 miles wide) lava-flooded, lagoon-like ring, is located on the southern shores of Mare Nubium. Its north wall is breached, and the walls can be difficult to trace in their entirety, being much eroded by the lava that spilled over their rims. The overall geometry of Pitatus's walls is very curious, with many undulations, including some wide, shallow, wriggling valleys on the east and southeast ramparts. Several nice craters lie atop the walls of Pitatus, one on the west wall having a prominent central peak. A broad valley leads through the west wall to the adjoining walled plain Hesiodus.

The floor of Pitatus is very smooth and dark, completely flooded by basaltic lava. There are exciting details to observe on the floor, however – it is far from featureless. Several small but interesting groupings of low mountains lie here and there inside the walls, especially to the south and west. There is a central mountain mass just north of dead center – it looks rather low, but you are just seeing the top of the mountain, the majority of it no doubt buried beneath the lava flow. Be sure to look for three white patches on the floor – two in the south (one southeast of the central peak, another due south of it) and one in the northwest corner. But the most striking of the floor features is the remarkable series of rilles that snake practically all the way around the inside walls. Only along the inside southern wall do these rilles disappear; they are most conspicuous along the northwest and northeast walls. Inside the northwest and west walls the rilles fork and zigzag; to the east they curve in a concentric way with a mountain group.

Hesiodus is a very interesting walled plain, though only 45 km (28 miles) wide. It is linked to its neighboring walled plain, Pitatus, by a wide shallow valley that breaches the west wall of the latter formation. Several gaps appear in the north wall of Hesiodus, which, like its neighbor, is lava-flooded. In place of a central peak Hesiodus has a perfect bowl crater. And, most amazingly, also like its neighbor, Hesiodus has a series of rilles running around the insides of its inner walls. These are easiest to see along the inside west wall, but I have seen them also to the north. Several small, fresh craterlets dimple the otherwise pristine lavas on the floor of Hesiodus.

The highlight here is actually **Hesiodus A**, the most striking example on the Moon's surface of a concentric ring crater. Touching the southwest wall of Hesiodus, it is a perfectly-formed bowl with a slightly offset secondary rim some way inside the outer walls, with a nice little group of central peaks at the very center of the concentric walls. Not to be confused with the subtle rille on the floor of the main crater is the **Hesiodus Rille**, a much longer and wider rille that begins outside the north rim and runs southwest, far across Mare Nubium, onto rough highlands and back down across Palus Epidemiarum.

Mare Nubium, the Sea of Clouds, is the largest mare area in the southwest quadrant, and also one of the "shallowest" of the lunar seas. How do we know this? Because, unlike many maria, which have only a few craters at their elevated shores where the lava flows couldn't cover them, you can see many craters even in the middle of Mare Nubium, though most of these are filled practically to their brims with lava, and many of their rims are barely visible, and hence are called ghost craters. It is thought that Mare Nubium is a very ancient impact basin, and it contains a good number of interesting objects, some of which we shall discuss below.

Longomontanus is an extensive walled plain, 145 km (90 miles) across, whose largely ruined walls are very heavily cratered. Wilkins and Moore describe the walls as "abounding in ridges and rocky spurs" and as containing "dykes of rock which stand out very prominently at times," but I am convinced that most of these landforms are also the rims of craters that intrude into the ramparts of Longomontanus. The south walls are very broken down and heavily cratered. The floor is at first glance unremarkable, but close inspection will reward the careful lunar observer. There is a very small central peak, and west of it a much larger mountain group capped with a summit crater. See if you can spot the several very small, but definitely elliptical craterlets that dot the floor – some of these are surprisingly elongated.

I have also noticed many small, round white spots on the floor of Longomontanus that are very different from the larger white patches on the floor of Pitatus. Some of the sizable craters that intrude into the walls of Longomontanus continue down onto the floor, especially in the northwest. Can you see the little rounded dome halfway between the central peak and the east wall? Between this dome and the wall is a buried crater ring that looks like a horse-shoe. As for the terrain outside Longomontanus, as Wilkins and Moore so aptly stated, "All around are many craters." This statement holds true for the southern lunar highlands in general – when drawing a feature in this region, it is strongly recommended that you confine your efforts to depicting only the feature itself, as any attempt to depict all of the surrounding craters will require too much time and effort.

Heinsius is a peculiar impact site, similar in some respects to Longomontanus, but at 72 km (45 miles) across only half the size. Like Longomontanus, Heinsius is overlapped by several large craters which interfere with its rim – these craters veritably overwhelm Heinsius itself, not an unusual situation in the heavily cratered southern highlands. The two principal overlapping craters are designated **Heinsius B** and **C**, and they are large enough to show details of their own; B has a central hill, while C has two small craterlets associated with it and a break in its southern wall. **Heinsius A** is another large crater, on the floor of the main walled plain, having its own central peak. Unlike Longomontanus, however, the inner walls of Heinsius show typical terracing. Smaller craters may be found on the crests of the southeast and northeast walls. Outside the walls are many ruined crater rings in all directions.

Wilhelm is another great broken-down impact site, 97 km (60 miles) in diameter, between Longomontanus and Heinsius. Like other such structures in the region, Wilhelm's walls have been disturbed by later impacts which created smaller craters that are so numerous that they have "weathered" the walls. Older lunar guidebooks attributed this appearance to "erosion," but it would be a mistake to make any comparison between the meteoritic bombardment that broke down Wilhelm's walls and the terrestrial process of erosion, which involves forces like weather, including wind and rain, and running water, none of which are factors in erosion on the Moon.

What remains of Wilhelm's walls are nonetheless quite impressive, soaring to nearly 3500 m (about 11 000 ft) above the floor below. Overall the walls are very irregular and distorted, squeezed into a sort of warped hexagon. Two very large craters impinge on the west wall. The floor is rough-hewn, with several bowl craters and rocky ridges. Just northeast of center is a very fine craterlet chain which I have mistaken for a rille. This is a common occurrence with lunar observing – when you are looking at a series of discrete objects like craterlets that are at the threshold of your resolving power, your eye and brain conspire to convince you that you are observing a linear feature; that is, you tend to "connect the dots." The illusion disappears under high magnification and conditions of excellent seeing. An elongated mountain is strangely positioned just inside

the southwest wall. A spike-shaped canyon penetrates the southeast wall, and points straight at a very odd crater known as **Montanari D**, which looks rather like one of the "soft" clocks in the Salvador Dali painting *The Persistence of Memory*. This remarkable feature has a bizarre serpentine ridge dividing its floor in half.

Clavius is one of the greatest of all lunar formations, 235 km (145 miles) in breadth and containing an incredible treasure-trove of details. Clavius has been described as a walled "depression" because the terrain around its rim is as high as the rim itself, except for parts of the east wall, which is an elevated ridge. The floor of Clavius has been described as convex. The classic guidebooks describe "terraced" walls, but I have never been able to make out more than a bare hint of terracing along the inner slopes of Clavius's walls, which are very steep and rugged cliffs. The fact is, Clavius is a very old impact site whose features have been "sandblasted" by meteoric bombardment, the action of micrometeorites, and perhaps certain geologic processes, so that features normally found in fresher, younger craters, like terraces, have been all but obliterated. You can certainly see some landslips along the inner northwest wall, but even here cratering has greatly confused the view. Wilkins and Moore describe the cliffs of Clavius as being "disturbed by depressions" – these depressions are almost certainly smaller impact craters of all sizes and shapes. The sheer rock faces of these cliffs rise an average of 3500 m (12 000 ft) above the floor of the formation, but they are also capped by occasional peaks that rise a further 1500 m (5000 ft) or so in places.

Perched atop the south wall of Clavius is the sizable (40 km, 25-mile wide) secondary crater **Rutherfurd**, formerly designated **Clavius A**. A curious aspect of this crater is its "central" peak, which is offset far to the north of the center, just inside the north rim! To me, it has often appeared that the base of this mountain mass is actually on the vertical rim, so that the peak sticks out sideways, precariously overhanging the floor of the crater. This is no doubt an optical illusion, due in part to the oblique angle at which we observe Clavius, which is subject to great foreshortening by its proximity to the Moon's southern limb. The north wall of Rutherfurd is very interesting – look for a series of long ridges which gradually slope northward onto the floor of Clavius. Elger likened these ridges to "the ribbed flanks" that adorn the great Java volcanoes. Abutting the southern rim of Rutherfurd are two distinct craterlets, and a gorge can be spotted cutting a swath across the southern floor where it meets the base of the wall.

The north wall of Clavius sports a twin to Rutherfurd, known as **Porter**. It also has a central mountain mass, with three peaks, but this mountain is indeed in the center of the crater. Another good-sized mountain is tucked inside the northwest wall of Porter, which is named after Russell W. Porter, the American telescope maker who was very involved with the founding of the famed Stellafane gathering of amateur astronomers (the oldest continuing event of its kind) and the design of the great 200-inch (5 m) Hale Telescope atop Mt Palomar, California. Look for a large four-sided depression adjacent to the west side of Porter, typical of depressions found all over the region. Both Rutherfurd and Porter take on a beautiful and dramatic appearance when the first rays of sunlight obliquely illuminate their rims – at this time, which occurs a day after first quarter, the rims look like white rings hovering over inky black pools, which of course are the craters' floors, still fully in shadow

The floor of Clavius is no less fascinating than the walls. A curious arc of four craters, each one slightly smaller than the last, crosses the floor in a roughly east–west attitude. These have been labeled **Clavius D**, **C**, **N**, and **J**, the first being the largest. If you include Rutherfurd in this group, you can extend the arc's membership to five craters of descending (or ascending, depending how you look at it!) size. The central peak is almost lost among these craters – it is not very impressive, to say the least. In fact, were you to stand atop this peak and look all around,

you would not be able to see the walls of Clavius, due as much to the Moon's pronounced curvature (which makes for very much shorter horizon distances than on the Earth) as to the modest elevation of your observation point. Besides those already mentioned, there are many smaller craters inside Clavius – I have counted more than two dozen during steady seeing conditions.

The region surrounding Clavius is uniformly scarred with an uncountable number of crater rings, large and small. Immediately outside the southwest wall lie two impressive walled plains, Blancanus and Scheiner, which are described later in this chapter. To the northeast is Maginus, which we shall visit next. To the east and west lie rugged highlands that can be explored for many a night by the ambitious lunar observer.

Figure 7.2 is a drawing of Clavius at sunrise. Rutherfurd, Porter, and Clavius D and C were filled with shadow, but you can see parts of their walls as white rings, as described above. The shadow cast by the west wall of the main crater was intensely black near the wall itself but this blackness softened as the shadow faded towards its boundary. At the upper right is the bright wall Blancanus.

Clavius is so large that you can see it as a dark notch in the southern terminator one day after first quarter (the same time to see the "white ring" effect with Rutherfurd and Porter). It is easy to imagine why Goodacre described Clavius as an object "of remarkable grandeur and absorbing interest." The

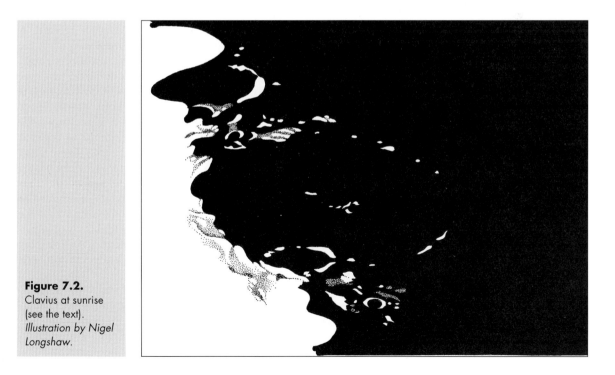

Figure 7.2.
Clavius at sunrise (see the text). *Illustration by Nigel Longshaw.*

hold that Clavius has on the lunar observer was perhaps best described four decades ago by Patrick Moore:

> So large is Clavius that, when on the terminator, it appears as a great bay filled with black shadow which perceptibly blunts the south cusp to the naked eye. Gradually the crest of the entire wall becomes visible, followed by fine rings of light, the walls of the large floor craters ... like looking down the funnels of a large steamer ... When the floor is revealed, the shadow of a high peak on the [east] wall is cast in striking steeple-like outline. The larger floor craters retain shadow when the rest of the floor is fully illuminated.

Maginus, a largely ruined crater whose structure is very similar in many ways to that of great Clavius, requires a little effort to make out its outline. At 180 km (110 miles) across, Maginus is huge, with 4250 m (14 000 ft) high cliffs overlooking its broad floor, and like Clavius, these cliffs are everywhere interrupted by a large variety of impact craters, some of considerable size. As with Clavius, the tops of the cliffs appear to be more or less level with the terrain that surrounds Maginus, so that it may also be considered a depression rather than a true walled plain.

Maginus C is the largest of the craters to disturb the walls of the principal formation; it is to be found atop the southwest bluffs. It is interesting that the even the rim of Maginus C is itself disrupted by cratering, with a conspicuous craterlet visible on the south rim and a very impressive chain of overlapping craterlets intruding from the northwest. Be sure to observe a nice grouping of craters that lie along the northwest boundary of Maginus, completely destroying it – these are designated **Maginus F, G**, and **N**.

The floor of Maginus contains a number of craters and craterlets, and a group of dune-like hills just east of center; otherwise its texture is relatively smooth. The most interesting parts of the floor are the margins, where the rims of several ruined or partly ruined crater rings that impinge upon the inner walls protrude down onto the floor, creating bay-like enclosures. Look for these structures especially inside the south wall. The area around Maginus is very similar to the surroundings of Clavius – to the north lies the crater **Proctor**, a ringed plain with strangely undulating walls.

Saussure, a 2500 m (8000 ft) deep, 50 km (30-mile) wide depression north of Maginus, is immediately south of, and tangential to, Orontius, a similar, but larger and even more ruined depression. There are many craterlets atop the crests of Saussure's cliffs, including three that are fairly easy to see in

amateur telescopes, at the northeast, west, and southwest rims. The south wall is most interesting, where several rough canyons have cut through it, some of them completely crossing the south rim, and others spilling onto the interior floor. To the east is a wide valley bounded by a mountain ridge topped with a most remarkable chain of craterlets. This ridge is concentric with the eastern wall of Saussure. The floor contains a few tiny craterlets, but no central peak.

Orontius is a large, squarish depression northeast of Tycho, whose walls are rather difficult to discern because they have suffered heavy cratering by subsequent impacts. The square shape is probably an illusion created by the intrusion of several large craters which surround it. Orontius has no central peaks, but its floor is pocked by many craterlets, several of which show curious trenches radiating from them in a northerly direction, giving them the appearance of miniature comets with tails. This suggests that a shower of impactors arrived from the south, digging out the craterlets, then skipping northward a bit further. The trenches are a little lighter in color than the craterlets themselves. Wilkins and Moore believed that the craterlets of Orontius were formed by secondary debris from the Tycho impact.

Nasireddin, a 50 km (30-mile) wide walled plain showing a surprising amount of detail for its size, lies east of Orontius and overlaps **Huggins** (which in turn intrudes into Orontius). Nasireddin's walls have been called "regular" by some of the classic lunar observers, but I find them anything but that. There is fine, undulating terracing all around. A great spur juts out from the southwest wall, while the northwest rampart is cut by a short, curving gorge. The eastern rim is sharp, like a knife-blade, while the western rim, where the crater overlaps Huggins, is much more rounded and broken down. Huge rounded hills, looking like giant sand dunes, lie inside the north wall, while the floor of the crater is broken by numerous hills and rugged ridges. There is a central craterlet on the floor; this feature has an irregular, elongated shape that is typical of other craterlets on the floor of Nasireddin – elongated in the north–south direction. Immediately north is the crater **Miller**, separated from Nasireddin by a horseshoe-shaped ring. Be sure to look also for a striking group of three large overlapping craters just outside the southeast rim of Nasireddin.

Walter is a huge and complex walled plain southeast of Orontius. This 145 km (90-mile) wide pentagon with highly disturbed walls that are as much as 3000 m (10 000 ft) above its no less tortured floor. Let's describe each side of the pentagon in turn, beginning with the south wall, which is little more than a series of hills and valleys. Moving clockwise around the pentagon, we next come to the southwest wall. Its inner slope is cut by a fine, dark line that is probably a landslip boundary; its outer wall is bounded by a zigzagging ridge that terminates in a sizable craterlet at its west end. Next comes the northwest wall, whose crest is about even with the ground outside the formation. Here is a fine low ring called **Walter W**. At the base of the northwest wall where it joins the southwest wall is a fine crater, known as **Walter E**. The northern half of the northwest wall is composed of two parallel crater chains that head towards a large triple crater on the northwest rim – **Walter A, B,** and **C**.

The north wall of Walter is characterized by a wavy gorge along its entire length, while the east wall, the final side of the pentagon, is dominated by another triple crater, the largest member designated **Walter D**. Also abutting the east wall is the even larger crater **Walter L**, and the still larger, mostly ruined 32 km (20-mile) wide formation called **Nonius**. The floor of Walter is pocked with many large craters, particularly in its northeastern quarter, many of them so close to one another that their rims merge to create bright hills and ridges.

Purbach, another great walled plain of 120 km (75 miles) width, lies along the eastern shore of Mare Nubium. The east walls appear highest, rising to altitudes of 2500 m (8000 ft) or so. The west walls are broader, but not as high; there is a wide shelf along most of the west side where a huge landslip has evidently taken place. Purbach's southern wall merges with the northern wall of Regiomontanus, a very similar structure south of it. The floor of Purbach, according to Wilkins and Moore, "swarms with detail." The principal floor feature is a central ridge, most of which is actually the eastern rim of **Purbach W**, a partly ruined crater ring. The rest of the ridge is a distinct elevated spur coming off the north rim of the ruined ring. From the northwest rim of this ring runs another, longer ridge that terminates at the south edge of a large crater poised atop the northwest rampart of the main walled plain; this is **Purbach G**. Yet another ridge, shaped rather like a corkscrew, traverses the southern wall, carrying on into Regiomontanus. All of these ridges are oriented approximately north–south. On the east half of the floor a weird trench curves north and south, while you can, under very steady skies, spot a fine rille meandering along the inside of the west wall. **Purbach A** is a sharp little crater just south of center.

The area around Purbach is worthy of many nights of careful exploration. We have already mentioned its southerly neighbor, **Regiomontanus**, an oval, mountain-walled plain that measures about 130 km (80 miles) from east to west, but only about 105 km (65 miles) from north to south, mainly because of the intrusion of Purbach into its northern wall. Much of the southern boundary of Regiomontanus consists of very rugged-looking mountains, some of which rise 2100 m (7000 ft) above the plains. This formation is noted for its central mountain, capped by a large crater that makes it look for all the world like a terrestrial volcano. Northeast of Purbach and just touching it is the crater Lacaille, while the crater Thebit is to the northwest, seemingly connected to Purbach by a long, bending ridge.

Thebit is a deep, 50 km (30-mile) wide crater located on the southeastern shore of Mare Nubium, directly east of the Straight Wall. It is a most interesting crater indeed, its floor scoured in numerous places by short, deep trenches. Many of these trenches appear to encircle the central part of the floor, which is devoid of any conspicuous central peak or crater. On the south side the trenches take on a zigzag pattern. An isolated mountain stands in the northeast corner of the floor. A large landslip is visible along a great length of the southwest wall; a fine set of terraces appears along the east wall. A large crater, **Thebit A**, perches atop the northwest rim; it has a central peak. Adjoining it is yet another crater, **Thebit L**, smaller still but also with a prominent central peak.

The **Straight Wall**, one of the Moon's most famous features, stretches along the eastern shore of Mare Nubium, and is best seen a day or two after first quarter. This feature, 130 km (80-mile) long, runs northeast–southwest, bisecting a "bay" in Nubium that is actually a ghost crater. This feature is not perfectly straight – you will easily be able to detect small undulations along its length, especially under high magnification – but overall it is remarkably linear, unlike anything else on the lunar surface.

You might think, from the shadow it casts toward the west, that the Straight Wall is a lofty cliff. The shadow is very broad and dark, even against the darkened mare basalts. This effect is the result of the east side of the formation being higher than its west side. But it is only 200 m (700 ft) or so higher, and detailed studies of the changing shadow as the Sun angle increases suggest that instead of being a steep cliff, the Straight Wall is basically a 40° sloped ridge. This is still fairly steep, but it is not a sheer cliff-face. The Straight Wall has been explained as a fault in the lunar surface, but its origin remains unknown. At the south end is a 580 m (1900 ft) high group of mountains called the Stag's Horn. Figure 7.3 is a drawing of sunset on the Straight Wall.

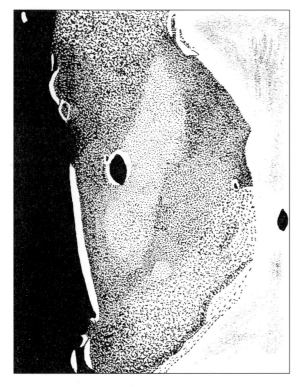

Figure 7.3. Sunset on the Straight Wall. *Illustration by Sally Beaumont.*

While you're in the neighborhood of the Straight Wall, be sure not to miss the 18 km (11-mile) wide bowl crater **Birt**, its smaller companion bowl crater **Birt A**, and the long sinuous rille to its west called **Rima Birt** that runs roughly parallel to the Straight Wall. See if you can spot the two craterlets that lie at either end of Rima Birt – the 3 km (2-mile) wide craterlet at the north end is **Birt E**, and the tinier craterlet at the south end is **Birt F**, only 1.5 km (1 mile) or so in diameter.

Ptolemaeus is near the center of the Moon's nearside, and along with its neighbors to the south, Alphonsus and Arzachel, it forms an impressive trio of large walled plains – at first quarter you simply can't miss these giant, impressive features. The group offers a chance to compare and contrast older and newer lunar features. Ptolemaeus is more than 145 km (90 miles) from rim to rim, and its very complicated walls, broken up by large numbers of valleys, canyons, and passes, betray the formation's old age.

It is hardly a fresh impact site – an awful lot has happened in and around Ptolemaeus since it was formed. Unlike fresher impact sites, there is no sign of terracing here – just discontinuous walls consist-ing of a jumble of mountains and ridges. And most of these "mountains" are not all that high, another sign of old age. Although the rim of Ptolemaeus has a peak that is 2750 m (9000 ft) high, that is the exception, not the rule, as most of the rim is no more than 1000 m (3000 ft) or so above the surrounding terrain. Much of the "broken" appearance of the walls of Ptolemaeus is the result of later impacts that have occurred, seemingly, everywhere.

The floor of Ptolemaeus is pretty interesting, too. A favorite game of selenographers who have studied this site over the years is "count the craterlets." They produced a series of charts, each purporting to show a greater number of craterlets than earlier efforts. In fact, I can understand what motivated them – so many of the craterlets on the floor of Ptolemaeus are just at the threshold of small- to moderate-sized telescopes, and appear to pop in and out of view, depending on the momentary steadiness (or unsteadiness!) of the sky. **Ptolemaeus A** (labeled **Lyot** on older maps) is by far the largest of them all, at a mere 7 km (4.5 miles) across. Drawing all of the craterlets you can see on the floor of Ptolemaeus can be a tricky business, as they are visible only at glancing Sun angles, lacking rims or other relief that would allow them to stand out more clearly from the floor itself. Back in the 1950s Hugh Percy Wilkins produced a particularly fine chart of the formation, depicting a breathtaking amount of detail. There is, however, no central peak (another sign of geologic senility), nor are there high ridges on the floor.

Alphonsus, immediately south of Ptolemaeus, is 110 km (70 miles) wide and obviously more youthful than its larger neighbor. How can we tell? Well, for one thing, it appears that the north rim of Alphonsus has superimposed itself upon the south wall of Ptolemaeus, suggesting that the impact creating Alphonsus happened later than the impact responsible for Ptolemaeus. Moreover, the features of Alphonsus are sharper and less degraded than those of Ptolemaeus, and it has some features, like a central peak, which Ptolemaeus lacks altogether. But Alphonsus is not as youthful as Arzachel, for its walls are not as high, they lack the highly developed terracing, and its central peak is not as well-preserved as Arzachel's. So Alphonsus not only is the "middle" crater geographically of the three, it is also intermediate in age.

You can spend many, many lunations studying the walls of Alphonsus, which rise to heights of 2000 m (7000 feet). They are heavily dimpled by depressions, many of them having an elliptical shape. Like Ptolemaeus, the walls of Alphonsus abound with valleys and passes. Look for the very long, wide valley that cuts clear through the southeast rampart; it is interrupted only once outside the crater's wall by a high ridge, then continues for a distance towards Arzachel. Other, similar gorges parallel to this valley are visible just outside the southeast wall of Alphonsus. Against this wall you will also see a large, degraded crater with a tiny central peak, called **Alphonsus D**; a curious aspect of this crater is that one valley enters it from the north wall, while a matching valley exits through the opposite (south) wall.

The floor of Alphonsus has long been the source of much excitement and controversy. Many observers have reported seeing lunar transient phenomena (LTPs) here, especially in conjunction with the central peak, which is probably the mere remnant of a more impressive mountain that has broken down over the ages. In November, 1958, the Soviet astrophysicist Nikolai Kozyrev focused his spectrograph on the central peak of Alphonsus and purportedly obtained a spectrum showing emission lines in the blue wavelengths, which he interpreted as proof of a gaseous emission from the peak. Kozyrev's results, which have been called into question and were never duplicated, are discussed more fully in the section on LTPs in Chapter 10.

Equally controversial are the dark spots that can be seen on the floor of Alphonsus. Three large dark spots are easily visible in amateur telescopes – one just inside the southeast wall, one just inside the northeast wall, and the third just inside the west wall. That all three are just inside the inner ramparts of Alphonsus is certainly curious and perhaps a clue to their nature and origin, but no one has ever fully explained them. At one time, some astronomers even speculated that the dark patches were vegetation! A more likely explanation is that they were produced by small-scale eruptions of lava, similar to the great lava flows that created the vast expanses of dark mare basalts, but on a much smaller, more localized scale. More diminutive dark spots are found elsewhere inside Alphonsus, including one southwest of the central peak.

Not very controversial but well worth observing is a system of delicate rilles that cross the eastern floor of Alphonsus from north to south. The easternmost rille, not far from the east wall and running parallel to it, is forked, and actually runs across the dark spot in the northeast corner of the floor. Another rille, even more delicate, lies west of the first rille, between it and the central peak. It is extremely difficult to see, even in large amateur telescopes, and requires very steady skies. Easier to see is a series of broad gorges and narrow ridges that run from the central peak to the south wall. Figure 7.4 is a

Figure 7.4. Sunrise on Alphonsus. *Illustration by Sally Beaumont.*

drawing of sunrise on Alphonsus, showing the central peak as "brilliant white," casting a long spike of a shadow that reaches all the way to the western rim of the crater.

Arzachel, the youngest of the trio of great walled plains near the eastern shore of Mare Nubium, has features typical of fresh impact sites – high, well-defined walls with extensive terracing, and a rugged, impressive central mountain mass. Arzachel is 96 km (60 miles) wide, and clearly has higher walls than the other members of the trio since it takes the rising Sun much longer to fully light up its floor than with the other two. To the west, the walls rise 3000 m (10 000 ft) above the floor, and the eastern walls are higher still – as much as 4100 m (13 500 ft). The terracing is obvious even to the novice lunar observer, especially along the western walls, where the terraces take on the aspect of great waves. The outermost rim of the southeastern wall is separated from the inner parts of this wall by a long, wide canyon.

The floor of Arzachel contains at least three features of great interest. The first is the central mountain mass, which appears as a north–south ridge 1500 m (4900 ft) high, with one craterlet just south of its summit and another at its southern foot. The second noteworthy feature is a rille that winds its way across the eastern half of the floor, from the north rim to the south rim. West of this rille is the third feature worth looking at, **Arzachel A**, a deep crater with a tiny peak, not at its center but just inside its southwest rim.

Alpetragius, nestled between Alphonsus and Arzachel, is a deep crater with a disproportionately large central mountain mass, and makes for an interesting telescopic object. Despite a diameter of only 43 km (27 miles), Alpetragius has walls that tower 3700 m (12 000 ft) above its floor. There is not much of the floor to see, however, for most of it is covered by the central mountains, which are 19 km (12 miles) across and 1800 m (6000 ft) high or more.

Lassell is an unassuming 23 km (14-mile) wide crater on the surface of Mare Nubium, with polygonal walls and a very dark floor that matches the mare basalts all around it. There is a depression in the north wall, and a few fugitive features on the floor, but little else. The region immediately outside this crater is interesting, as it contains several splotches of bright material that contrast well with the dark mare background.

Mösting, a rather eroded, 24 km (15-mile) wide ring plain with a depressed floor, has outer walls a mere 500 m (1600 ft) above the ground outside the crater, but fully 2100 m (7000 ft) above the crater's floor. To the south is the tiny **Mösting A**, the center of a modest ray system.

Herschel, on the northern border of Ptolemaeus, and similar in many ways to Agrippa and Godin, is a 37 km (23-mile) wide ringed plain with highly terraced, very circular walls, very bright on the interior. Instead of a simple, single central peak, Herschel has a complex of peaks, perhaps as many as four, at least one of which is graced by a central craterlet. This feature was discovered by the selenographer Julius Schmidt. To the north of the main crater lies **Sporer**, a fairly intact walled plain that nevertheless shows some filling in with lava. Very noteworthy is a huge rugged gorge in the shape of a cigar, 130 km (80 miles) long, to the east of Herschel.

Flammarion, between Herschel A and Mösting A, used to be considered part of the Herschel system of craters. You will see this feature as an expansive and open, irregularly shaped ringed plain, bordered here and there by various mountain ridges. It is so poorly defined that Goodacre criticized it as "quite unsuitable for a separate name." It may therefore be regarded as a mild observing challenge to pick out. In the center of this crater is a shallow bowl-like depression, similar to those seen in Ptolemaeus, with yet another one of these depressions seen in the northwest quadrant. The entire floor is riddled with craterlets. Look for the crater **Flammarion K** on the north edge, and a rille running from it toward Lalande. West of Flammarion K this very same rille branches out, the offspring heading for the center of Flammarion itself. On the extreme south side you should be able to detect another craterlet, **Flammarion I**, with a rille that courses through a break in its southeast wall.

Bullialdus is certainly one of the grandest of all the walled plains on the Moon's surface. At 63 km (39 miles) in diameter with massive, 2500 m (8000 ft) high, roughly terraced walls and a huge blanket of bright ejecta debris, it looks every bit like a smaller version of Copernicus. Situated right in the middle of the Mare Nubium, Bullialdus has no competition from other features in the southwest quadrant of the Moon. The great number of landslips inside its walls tend to give parts of those walls a squared-off appearance. The southeast rim is bent into a wavy form. The terracing did not produce perfectly smooth ledges, however, as the inner slopes are dotted everywhere with large outcroppings of rock. The dustings of bright ejecta on the outer slopes make it easier to see the subtle "ridges" that radiate down them to the mare plain below. I am not so sure that these are really ridges at all, despite the classical lunar guidebooks labeling them as such, for they have always appeared to me more like shallow trenches or chains of minute craterlets carved by debris ejected from Bullialdus during the impact that produced the crater. Some appear gently formed, like the path a rock might leave as it tumbles down a gentle slope.

A fine grouping of central mountain peaks dominates the floor of Bullialdus, the highest rising over 1000 m (3000 ft). A rocky spur juts from the southeast wall toward these mountains, and almost reaches them; they are also approached by a number of smaller hills that swarm all around, covering most of the crater's floor. There are several attractive craters of different varieties surrounding Bullialdus on the Nubium plains. Just outside the southeast wall is **Bullialdus A**, which Wilkins and Moore saw as connected to the main crater by a wide valley. Oddly, the southeast wall of this crater mirrors the southeast wall of Bullialdus, wiggle for wiggle! Due south of the main formation is **Bullialdus B**, a deeper crater that looks like a miniature version of Bullialdus itself, right down to its group of four central peaks and distorted, terraced walls. Much farther from Bullialdus, to the southwest, is the crater **König**, 23 km (14 miles) wide, and also with polygonal walls. To the northwest we encounter **Lubiniezky**, a 39 km (24-mile) wide ring that barely rises above the lava floods of Nubium, which have erased all of its interior details. You can trace the walls of Lubiniezky for almost a full 360°, but there is a definite gap in the southeast rim.

Guericke is a magnificently detailed, 58 km (36-mile) wide crater northeast of Bullialdus jutting out onto the Mare Nubium. Look for gaps in the south and east walls; an 850 m (2800 ft) high peak graces the north wall. A highly detailed chart made by observing this feature with the great 0.83 m (33-inch) Meudon refractor appears in Wilkins and Moore's classic guidebook – many of the details noted were unknown previously. **Guericke A** is a nice round crater not far inside the main formation's southwest wall; there is evidence of a discrete rille emerging from Guericke A, penetrating the south wall, then meandering its way through the countryside south of the main crater. Can you see it?

Many smaller craterlets pock the complex floor of Guericke, which shows uneven shades of light and dark that may be real albedo features, not just the result of different Sun angles. Look for a series of radiating wrinkle ridges running northward from Guericke's north wall. Figure 7.5 beautifully illustrates Guericke and the regions immediately north and south of it.

Campanus and Mercator are a fine pair a walled plains, both a little less than 50 km (30 miles) across, located on the southwest shore of Mare Nubium. It makes an interesting exercise to compare and contrast the two structures. **Campanus** has sharply defined, 1800 m (6000 ft) high walls that show much

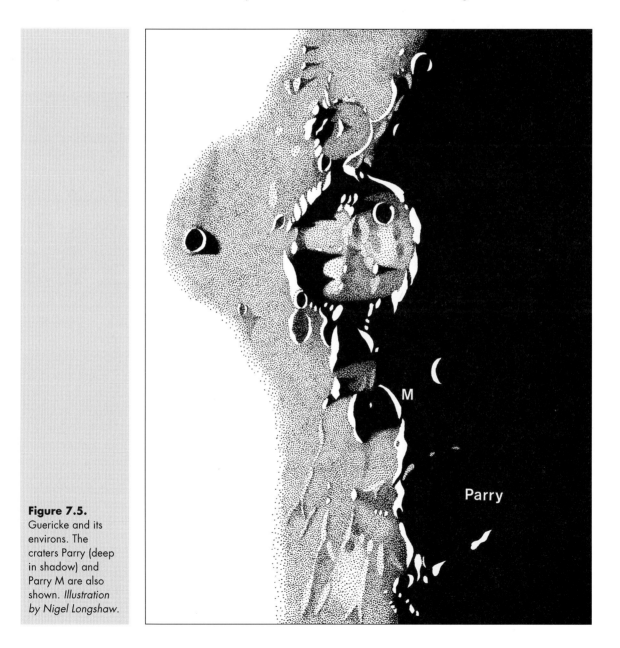

Figure 7.5.
Guericke and its environs. The craters Parry (deep in shadow) and Parry M are also shown. *Illustration by Nigel Longshaw.*

terracing, especially on the inner western slopes. Some of the peaks on the eastern rampart are quite jagged – look at their shadows. The crests of the rims are razor-sharp. On the rather dark floor you will see an arc-like central ridge. Inside the east wall is a rugged hill, and three bowl craters dimple the floor's north half, with a fourth just inside the south wall.

Mercator, on the other hand, has walls that are broad and somewhat eroded, with no sharp rims. Two fine craters rest atop the east wall, and a third crater sits atop the west wall. The north wall is missing a large piece, where it opens onto a broad lava valley that separates Mercator from Campanus – it is easy to imagine the lava spilling into Mercator through this gap in the wall. The lava flooding seems more thorough in Mercator than in Campanus, as it has no central peak or other features of note on its interior floor. An interesting group of hills may be observed just northeast of these two craters, which are depicted in Fig. 7.6, along with the ruined formation Kies, described a little later in this chapter.

Cichus is a 32 km (20-mile) wide crater located on the eastern shore of a small dark bay in **Palus Epidemiarum**, noted for the perfectly round, 8 km (5-mile) wide bowl crater, **Cichus C**, that rests atop its southwest wall. Look for the finger-like promontory that juts northward into the bay 40 km (25 miles) west of Cichus. The east and west walls of the main crater are broadly terraced, rising to 2500 m (8000 ft) on the east, 2750 m (9000 ft) on the west. A number of hills are scattered across the floor, and you should have no trouble spotting a curved rille that runs down to the floor from the southeastern wall. Outside the walls, to the northeast, lies an old ruined crater ring containing the

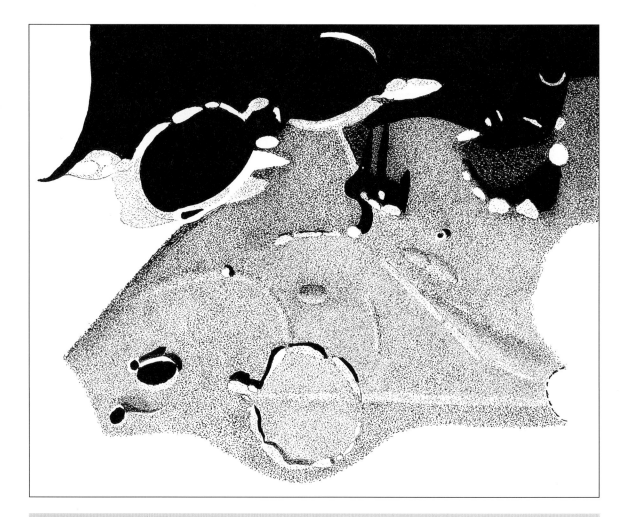

Figure 7.6. Campanus (top center) and Mercator (top left), both deep in shadow, and the ruined formation Kies (bottom center). Just above Kies is the dome known as Kies π. *Illustration by Colin Ebdon.*

little crater **Hauet**, with its sharp rim and nearly featureless floor.

Capuanus, a lava-flooded crater on the southern shores of Palus Epidemiarum, between Cichus and Ramsden, is 56 km (35 miles) across, with rims, which are 2500 m (8000 ft) high in places. The classic lunar guidebooks mention domes on the floor of this formation, but, except for a couple of bright hills tucked just inside the southwest and southeast walls, I have never seen any. One observer described the domes as being arranged in a "quadrangle" – there is a group of tiny craterlets having such a geometry, so perhaps these craterlets were mistaken for domes when seen at the threshold of resolution. Besides the craterlets, the floor is free of topographical features, though I have seen whitish streaks across it. Two large craters, the largest designated **Capuanus A**, sit atop the east rim, while a third (**Capuanus B**) rests inside the southwest wall. Adjoining Capuanus's southwest wall is a very elliptical, lava-flooded ring known as **Capuanus P**. The interior of this ring is very dark. From the west wall of Capuanus radiate two long mountainous ridges whose bright crests contrast sharply with the very dark mare surface of the Palus Epidemiarum.

Ramsden, an otherwise unassuming 19 km (12-mile) wide crater ring in the southwest part of Palus Epidemiarum, is famous for the extensive rille system that surrounds it. Two of the rilles begin at a point well northwest of Ramsden, in the highlands that separate Palus Epidemiarum from Mare Humorum. A single rille runs southeast from this point, then branches just northwest of Ramsden, one rille running along the southwest wall of the crater all the way to **Ramsden G**, a craterlet far south of Ramsden, the other rille running alongside the northeast wall almost all the way to the crater **Elger**. This forked rille intersects a second forked rille a little to the east of Ramsden. Yet another rille begins at a point between the tiny craters **Marth** and **Dunthorne**, and heads northwest to Mercator and Campanus.

Euclides is a tiny bowl crater, only 11 km (7 miles) across, and only 600 m (2000 ft) deep, beyond the western slopes of the **Riphaen Mountains** which border Mare Cognitum. Why bother to mention such a crater? Because it is surrounded by a very bright triangular nimbus of ejecta that shines with such a brilliancy when the Sun is high that Euclides is one of the brightest spots on the Moon's surface and, despite its diminutive nature, easy to find at full moon.

Lansberg, north of the Riphaen Mountains at the extreme eastern shores of Oceanus Procellarum, is a well-structured walled plain of 47 km (29 miles) width, well elevated above the dark mare basalts by its bright, ejecta-splashed outer slopes. There is much detail to observe and sketch along the outer walls, including many rocky protrusions. The 3000 m (10 000 ft) high rims are far from perfectly round. The inner walls show extensive terracing, and appear more gradually sloped than the outer walls, which are especially steep on the east side. Some of the terraces are not straight, but follow the curve of the crater, unlike the outer rims, which are squarish in many places. A fine group of peaks graces the center floor of Lansberg; a tiny triangle-shaped valley lies at their heart. Outside the main crater, to the southwest, you will find **Lansberg C**, a largely submerged ring-plain roughly half the size of its "parent" whose perfectly smooth floor testifies to the lava flows that spilled over its walls.

Flamsteed is a bright 14 km (9-mile) diameter crater on the western side of Oceanus Procellarum. Situated just inside the barely traceable boundary of a very large and ancient ring (**Flamsteed P**) whose walls are just a bare suggestion amid the extensive lava flows that occurred in the region, Flamsteed has irregular walls, a group of central hills, and some bright ridges across its floor. The ancient ring, 100 km (60 miles) across, encloses three bright bowl craters that are much smaller than Flamsteed. Nowhere do its "walls" rise more than 350 m (1100 ft) above the Procellarum lavas, and in some places only 50 m (150 ft) high hills are all that mark the rim's remnants. Flamsteed P is prominent at full moon, when its ring appears complete. The first Surveyor spacecraft landed here.

Letronne is an ancient lava-flooded ring whose north wall has been replaced by a wrinkle ridge, not an altogether uncommon occurrence on the lunar landscape. Because the north wall is missing, Letronne is classified as a "bay" lying at the edge of Oceanus Procellarum. Its floor contains the summits of three central peaks that lie mostly buried under the intruding lavas, but little else, the lavas having erased any other features that would have been visible many eons ago. Inside the eastern wall is a fine bowl crater, with a smaller one beside it. An almost completely buried ring intrudes into the western wall – this ring is called **Letronne P**.

Mare Humorum the Sea of Moisture, is a nearly circular lava plain that is somewhat foreshortened into an ellipse by its relative closeness to the Moon's southwestern limb. Although it's one of the smaller lunar seas, Mare Humorum nevertheless covers a lot of ground – 130 000 km² (50 000 square miles) – so it encompasses some interesting objects. It is hard to decide just where to begin when observing this region of the Moon. Do you start with the many mighty walled plains, like Gassendi, that sharply define Humorum's shores, or with the mysterious wrinkle ridges and many rilles superimposed on the lava plains? It really doesn't matter, for wherever you look you will find something to fascinate you.

If you look at Mare Humorum in a pair of binoculars, so that north is "up", you will see a dark oval patch of lava basalts ringed by rugged mountains to the west and south, with lower, more broken and isolated highlands forming the east boundary. All around are walled plains, of which Gassendi at the "top" (north) is by far the largest and most sharply defined. Along the southern shore are **Doppelmayer** and **Vitello**, while the eastern shores are broken by the mostly ruined and ill-defined rings known as **Hippalus** and **Agatharchides**. A little beyond the rugged western mountains lie several walled plains, the most prominent of which, Mersenius, is described below.

The floor of Mare Humorum is perhaps its most interesting aspect. Like all typical lava plains, most of the floor is a glassy-smooth surface pocked only by a few fresh bowl craters here, the lava flooding having erased all evidence of older structures. Some observers have identified two dozen of these bowl craters, but none are very large. If you look carefully you will see a couple of ghost crater rings that the lava flows have almost, but not quite, completely covered. Using either binoculars or a telescope at low power, look next at the eastern half of the mare floor, and you will notice several wrinkle ridges that meander north and south, roughly concentric with the east rim. Several different theories and models have been advanced by planetary geologists to explain the origin of these mysterious features, from subsidence to upwelling of lavas, but we may never know for sure which, if any, of these explanations is the correct one.

Interestingly, the wrinkle ridges on the eastern floor of Mare Humorum mirror a system of north–south rilles that are visible on the western floor, the most conspicuous of which is known as the **Doppelmayer Rille** for its proximity to the walled plain of the same name. A much more vast system of rilles winds its way through the highlands that make up the eastern rim of Mare Humorum, some of which may be traced for some 300 km (200 miles). These rilles are also oriented north–south. Did the same geologic process that produced these rilles also produce the wrinkle ridges that seemingly parallel them? Obviously, we still have much to learn about the Moon.

But by far the most dramatic object on the smooth floor of Mare Humorum is the 2100 m (7000 ft) high mountain range known as **Kelvin Promontory** (also called **Cape Kelvin**). Its elongated form is capped by two peaks that catch the first rays of sunlight streaming over the walls of Mare Humorum before any light reaches the sheer rocky faces of the range; at these moments, the peaks appear as brilliant white spots suspended over the dark mare floor below – it is quite a sight to behold! And last, but not least, at the northwestern shore of Mare Humorum you will find the partly-ruined crater **Kies**, which is best known for its associated volcanic dome called **Kies** π. Topped by a tiny crater pit, Kies π is the best example of a lunar dome you can ever hope to observe (see Fig. 7.6).

One look at **Gassendi** will tell you why it has been a favorite target for lunar observers for over two centuries. You will never tire of looking at this great, 88 km (55-mile) wide walled plain, for, like revisiting a masterpiece painting, each time you return to it you will almost certainly discovery something you hadn't noticed before.

The walls of Gassendi are worthy of your careful attention. To the east and west they are very high, containing peaks that have altitudes of 2750 m (9000 ft) or more. They are not very broad, which adds to the impression you get when observing them that they are very precipitous. Gassendi's walls have always reminded me of looking down at the top of a tin can which someone has opened with a rather old and dull can-opener, leaving irregular, jagged, razor-sharp edges. The east wall is almost perfectly straight, as is a shorter segment of the northeast wall, but the south wall is nicely rounded, as is the west wall, except for one peculiar place just south of due west where it is S-shaped. Obvious landslips appear inside the west and northwest walls, but nowhere else. The inside of the south wall is flooded with lava. The wall here is often no greater than 150 m (500 ft) above the floor. The north wall is broken by the intruder known as **Gassendi A**, a finely detailed walled plain that is younger and better preserved than its "parent" crater. Gassendi A, whose location gives the whole Gassendi formation a "diamond ring" appearance, shows extensive terracing and an extremely rough floor, covered by hills and deep gouges in its terrain.

As fascinating as the walls of Gassendi are, the real highlight of this feature is its vast and highly detailed floor, best known for the marvelous network of rilles that crisscross it in every direction. Mapping these rilles is great sport, and a good way of testing your skills as an observer and an artist. You will want to begin observing these rilles when the Moon is 11 days old – like all rilles, the rilles of Gassendi are subtle features that require low Sun angles to be seen to their best advantage. Most of the rilles are found on the eastern half of the floor. One rille begins just inside the south wall and runs northeast until it meets a ridge that continues in the same direction towards the east wall. Where it meets the ridge this rille makes a sharp left-hand turn, heading north past central peaks (described below), and trickling to a stop in some rough hills in the far northern reaches of the crater. This rille has another branch to the west of the first that terminates at the central peaks. Other rilles wend their way through the eastern hill country of Gassendi, while another scours the surface northwest of the central peaks. Still others are clearly visible southwest of these peaks.

Besides the rilles, the other distinguishing characteristic of Gassendi's floor is its roughness – everywhere there are little pointy hills, and the central floor is dominated by a very impressive group of mountains arranged in a horseshoe pattern. The highest peak in this group is about 1200 m (4000 ft) in altitude. Just east of these mountains I have detected a large, shallow, almost perfectly-round bowl crater, but this is not shown on any lunar map I have consulted. It is, however, shown quite clearly on a Lunar Orbiter photograph, so I know that, in this one instance anyway, my imagination did not get the better of me! I think it is interesting to realize that the north floor of Gassendi, which rests atop the highlands at the northwest shore of Mare Humorum separating Gassendi from Letronne, is 600 m (2000 ft) higher than the lava plains of Mare Humorum itself. The south half of the floor is lower, and lavas have clearly intruded from Humorum through at least two wide gaps in the south wall, smoothing over any details that might once have been visible there. Figure 7.7 shows Gassendi at sunset.

Mersenius, west of Mare Humorum, is a great walled plain of 72 km (45 miles) breadth and 2100 m (7000 ft) high finely-terraced walls, associated with a fine system of rilles that parallels the western shores of the mare. The broad walls are very rugged and cut by many valleys on all sides. Two especially deep parallel canyons cut through the southwest wall, and the south wall is broken by a very wide but shallower valley. A large crater (**Mersenius H**) disturbs the southwest wall, next to the two deep canyons. A concentric ridge appears inside the entire length of the east wall, but the west wall has only numerous rounded hills inside it.

The convex floor of Mersenius will reward the patient observer with many fine details, including a most remarkable chain of craters perfectly aligned in a north–south direction. There is a dome-shaped central hill, and, like Gassendi, the floor of Mersenius is host to its own system of rilles, though not nearly as conspicuous. One rille parallels the

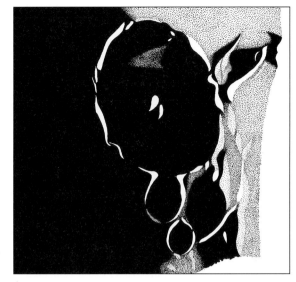

Figure 7.7.   Gassendi at sunset. *Illustration by Nigel Longshaw.*

crater chain, while others run east – west across the northern and southern floor. Outside the walls of Mersenius there is more of the same, but on a much larger scale, with the country here cut by huge rilles that are dozens of kilometers long.

Hainzel is a complex "compound" formation, the result of multiple impacts that have overlapped one another, making for a most confused scene. The overall shape is something like a pear. The northwest section of this formation has actually been designated **Hainzel A** – it is a fairly fresh, well-preserved crater 58 km (36 miles) in diameter. Hainzel A displays broad walls with much terracing, and typical central mountains. A series of ridges that are no doubt part of the terraces almost enclose the central peaks. The rest of Hainzel consists of a larger, older ring southeast of Hainzel A and overlapped by it, so that this older ring has no northwest wall. The older ring, whose southeast wall is interrupted in places by smaller craters, contains within it several curved ridges that are clearly the remnants of other impacts – one series of ridges marks the boundaries of the very irregularly shaped **Hainzel C**. To the north of Hainzel, look for a row of four similarly sized craters aligned east–west, all a little less than 16 km (10 miles) wide. To the south, you will see at least half a dozen more craters of similar dimensions.

Schickard, at 216 km (134 miles) across, enjoys the distinction of being one of the largest of the Moon's walled plains. Its walls are low – 1400 m (4500 ft) or so – though there are peaks on the south rim that exceed 2500 m (8000 feet), and ones on the east rim of more than 2900 m (9500 ft). Many craters surround the walls, and several lie atop the ramparts themselves. The walls are dimpled by numerous depressions which are probably eroded impact sites; there is no terracing detectable, further evidence that Schickard is an old formation.

The floor of Schickard is lava-flooded, and so, except for a few fresh craterlets, much of the interior is smooth and devoid of relief. A number of sizable craters do appear at the margins of the floor, where the lava was not quite as deep. Although smooth, the floor of Schickard is not entirely homogeneous, for its southern and eastern parts appear to be splashed with bright ejecta, contrasting them with the rest of the floor, which is very dark. The southern half of the floor is rougher and higher than the northern half, so it contains most of the larger craters. The southern floor also displays a series of parallel grooves that cut east and west.

Wargentin is a walled plain which appears to be completely filled to its 425 m (1400 ft) high brim by lava flows that give the whole formation the appearance of a mesa or bluff, a perspective that is enhanced by this feature's proximity to the lunar limb. Lavas have so completely filled this crater that the only feature visible inside it is a forked wrinkle ridge that runs down its major axis, looking like the tongue of a serpent. The walls of Wargentin probably do protrude above its floor by some considerable height, but from the oblique angle that we view this feature, that height appears negligible. Wargentin is about 90 km (55 miles) wide and very elliptical in appearance, but this of course is a foreshortening resulting from its location – Wargentin would appear circular if located near the center of the Moon's nearside face. Wargentin forms an attractive group with two other crater rings with which it shares its walls – **Nasmyth** and **Phocylides**. These two formations are not nearly as filled to their brims as is Wargentin, so their inner walls are visible and show considerable vertical relief.

If you can't make out the ancient crater **Bailly**, located on the extreme southwest limb of the Moon, don't feel too disappointed! The fact of the matter is, though it is the largest of the Moon's walled plains at 295 km (183 miles) across, Bailly is a very difficult object to delineate, for two reasons. First, its chaotic highland borders, which rise to heights of 4250 m (14 000 ft), have been very heavily eroded by all manner of smaller impacts, breaking its "walls" into so many fragments as to make it unrecognizable. Second, Bailly's location at the lunar limb places its far borders beyond our view. The interior of Bailly, assuming you can differentiate it from the exterior, contains two very large craters designated **Bailly A** and **B** (the latter being the largest) and many smaller craters, but no central peak and very few hilly areas. Like all objects perched on the lunar limbs, you will want to take advantage of favorable librations in longitude, which "rock" the Moon in an east–west direction, bringing Bailly more fully into view from our vantage point.

The **Dörfel Mountains** are a great mountain range located on the Moon's extreme southwest limb, with peaks rising more than 6000 m (20 000 ft). At least two peaks in this range attain altitudes of 8000 m (26 000 ft).

Kircher is a crater northeast of Bailly, having very high walls that rise 5500 m (18 000 ft) above its depressed floor. Amazing numbers of craters may be found adjoining and nearby Kircher, which is typical in all respects of the formations found in the Moon's rugged southwestern highlands.

Casatus, a 110 km (70-mile) wide walled plain, has unusually lofty walls – as high as 5500 m (18 000 ft) on the east and 6500 m (22 000 ft) on the west. A crater chain runs along the crest of the east wall, described by Wilkins and Moore as "a mighty line of white cliffs." Inside the southwest wall is a group of isolated mountains. The rims abound with large and small craters. Looking beyond Casatus towards the very edge of the Moon, you may be able to make out two very high mountain peaks gleaming against the black of space, when libration conditions favor it. Casatus adjoins the similar crater ring known as **Klaproth**, a 100 km (60-mile) wide walled plain that is shallower than its neighbor; its ramparts are also heavily cratered. The inner south slope of Klaproth is extremely rugged, with many ridges and rocky spurs than run down to the crater's floor. Two similar ridges protrude from the north wall. The center of the floor also contains a series of ridges.

Newton is often considered to be the deepest of the lunar craters, with walls that tower up to 7500 m (24 000 ft) above its interior. But because it is located so close to the south pole, it is not an easy object to observe, and it is very difficult to see any detail inside it. The lofty walls are partly to blame for this problem, as they keep the floor in deep shadow virtually all of the time. Newton's irregular pear-shape probably indicates that it is a multiple-impact site. It is perhaps 120 km (75 miles) across, east to west, but somewhat longer from north to south.

The **Leibnitz Mountains** are a great mountain range centered on the Moon's south pole. Even under favorable libration conditions the Leibnitz Mountains can be difficult to observe, and determinations of the heights of individual peaks in this range, which have been estimated to be as great as 10 000 m (33 000 ft), are to be regarded as somewhat suspect. The best time to view this range is when the Moon is three to four-days old, when the highest peaks are illuminated by earthshine. At times, some of the Leibnitz peaks appear as disconnected points of light above the cusp of the lunar crescent.

Blancanus, just southwest of mighty Clavius, is fine depression of 92 km (57 miles) breadth whose broad walls, though not highly elevated above the surrounding terrain, nonetheless manage to rise an impressive 3500 m (12 000 ft) above its floor. Blancanus has very broad inner slopes and rugged outer ramparts, all of which are disturbed here and there by cratering. It has a typical central mountain mass, also surrounded by several craters.

Scheiner, a 110 km (70-mile) wide walled plain, has richly terraced walls capped by peaks as high as 5500 m (18 000 ft) and is located near the great formation known as Clavius. The interior contains many craters and craterlets. There is no central peak, but you will might able to detect a few hills.

Moretus. A truly impressive crater, 120 km (75 miles) wide, with massive, complexly terraced walls, and a huge, 2100 m (7000 ft) high central mountain. Located southeast of Clavius, near the south pole, Moretus has every characteristic of a well-formed lunar impact site. The overall shape of the walls is circular, but they are warped outward to the northeast and southwest, and there are extensive landslips everywhere around them. Rocky protrusions are superimposed upon many of the ledges of the inner walls. These have been described as foothills. The central peak, which is said to be the highest structure of its kind on the Moon, is very steep, and there are several lower hills gathered around its base. The landscape around the outside of Moretus is nothing more than a gigantic crater-field, with impact sites of every imaginable size and depth.

Gruemberger, a 93 km (58-mile) wide walled plain, is located very close to Moretus. It has heavily eroded elliptical walls, and a very rough interior, covered by numerous hills and ridges. Gruemberger shares its northeast wall with the smaller crater **Cysatus**. There is no central peak on the rugged interior, but a very round crater designated **Gruemberger A** covers much of the southwestern floor.

Zupus is a little 19 km (12-mile) wide crater known for its extremely dark floor, and is about halfway between Gassendi and the Moon's western limb. Because its walls are low, Zupus probably owes the darkness of its floor to filling by lava basalts, somewhat like a miniature Wargentin. It

appears very dark even under high Sun angles. Classical astronomers thought they saw changes in the floor's hue, and one even ascribed these changes to the growth of vegetation, which we now know is, of course, impossible.

Grimaldi, famous as one of the darkest spots on the Moon, is a 195 km (120 miles) diameter lava-flooded walled plain just beyond the western shore-line of Oceanus Procellarum. It is bordered all around by innumerable individual mountains and cliffs, and its walls are nowhere close to continuous. Most of the bordering mountains are no more than 1200 m (4000 ft) high, but there is at least one peak that attains the lofty altitude of 2750 m (9000 ft). Because the walls are made up of so many separate highland fragments, it is virtually impossible to accurately portray them in a large-scale drawing. The west wall is marked by the craters **Grimaldi A** and **H**; **Grimaldi B** is a fine crater that interrupts the north "wall." The floor of Grimaldi is as smooth as glass, as dark as ebony, and as featureless as can be. It is interesting that the floor actually appears much darker than the lavas of nearby Oceanus Procellarum; it is so dark that with binoculars you can see Grimaldi by the light of earthshine when the Moon is four or five days old!

Riccioli is a huge walled plain at least 160 km (100 miles) across, foreshortened by its nearness to the Moon's west limb. The walls of Riccioli are very eroded and nowhere higher than 1200 m (4000 ft). The innumerable ravines and valleys that cut through the walls tend to be very narrow on the eastern ramparts, very wide to the south and west. Touching the west wall is a crater, called **Riccioli K**, with heavily eroded walls. A magnificent valley opens into Riccioli K from the north. Look for a very bright, corkscrew-shaped mountainous ridge just inside the northwest wall of the main crater.

The north half of the floor is characterized by an oddly shaped patch of very dark material (probably basalts) that contrasts with the much brighter southern half of the floor. South of the dark patch is a central ridge containing three bright peaks. Several unusual rilles cross the floor of Riccioli. Since the whole area around this crater abounds in rilles, we should not be surprised to find them inside Riccioli. But unlike the delicate, threadlike rilles we are used to seeing in the vicinity, those found inside Riccioli are very wide and shallow, looking for all the world like tank tracks across desert sands. One of these rilles crosses the patch of dark material and is almost swallowed up by it. Another starts far outside Riccioli and crosses, unimpeded, over its northwest rampart – this rille *is* swallowed by the dark patch.

The landscape around Riccioli is most remarkable. Besides the rilles, which we have already mentioned, there are endless series of rounded hills, and rugged ridges. One particularly amazing formation is the rectangular enclosure known as **Hartwig** which abuts the southwest wall of Riccioli. Massive canyons spill down the sheer rock face of Riccioli's walls into Hartwig, where they empty into a depressed area. The depressed floor of Hartwig is heavily scoured by at least two dozen trenches that run its length in a northeast–southwest direction. Close inspection of Riccioli's floor will reveal the existence of dozens more of these mysterious scourings, all running in the same northeast–southwest direction. In fact, the whole area surrounding Riccioli for hundreds of square kilometers, is inundated with these trenches, all having the same orientation. To the northeast is a beautiful example of a concentric ring crater, labeled **Riccioli C/CA**.

The **Cordillera Mountains** form a very lengthy mountain chain lying close to the western lunar limb, the principal peaks lying to the west of the crater **Crüger.**

The **D'Alembert Mountains**. are a range of mountains with 6000 m (20 000 ft) high peaks, located on the Moon's west limb, near Riccioli. Look for the great peak known as Table Mountain.

The **Rook Mountains** are another mountain range located on the Moon's extreme western limb and are therefore poorly placed for easy observation. Some of the peaks may also attain altitudes of 6000 m (20 000 ft) or more.

Byrgius is a 64 km (40-mile) wide, 2100 m (7000 ft) deep crater near the Moon's western limb. The eastern rim of Byrgius is crowned by a very bright, perfectly round crater called **Byrgius A** (also known as **La Paz**). Opposite Byrgius A, which is the center of a small system of rays, is **Byrgius D**, tangential to the western rim. This crater shows terracing and central mountains. The northern half of the floor of Byrgius is relatively smooth, but the southern half has curving rocky ridges and at least three craterlets. Be sure to notice the wide valley that breaks through the north wall of Byrgius and leads straight to the great **Sirsalis Rille**. Outside Byrgius, to the south, lies a large walled enclosure that looks like a giant kidney bean.

Chapter 8

Introduction to Observing the Moon

Now that we have surveyed the Moon to discover the most interesting surface features that can be seen or imaged with telescopes typically possessed by amateur astronomers, we are ready to get started with some real, hands-on observing of the Earth's satellite. But how and where to begin? What equipment will be needed? Where is the best place to set up the telescope for observing the Moon? Besides just looking at the Moon, can we make any meaningful observations? And do we need to have any specific skills to get the most from our lunar observing sessions? Finally, once we've made our observations, what can we do with them? These are all good questions, and are asked often by beginning lunar observers. In this chapter I shall attempt to answer these questions clearly and concisely.

Why Observe the Moon?

First off, it may be safely said that the Moon is almost always the very first object to catch the eye of the beginning amateur astronomer, for the simple reason that, being by far the biggest and brightest object in the night sky, it is also by far the easiest object to find. If your first telescope is a typical small refractor, of, say, 60 mm (2$\frac{1}{2}$ inches) aperture (the "department store telescope"), the Moon may be the only object you can easily locate at first, and it will show much more detail than anything else in the sky. Rare indeed are those who do not gasp with wonder and excitement at their first view of the Moon's craters and mountain chains through a telescope – any telescope. It is easy to see, then, why so many amateur astronomers aim their first tele-

scopes at our nearest celestial neighbor. Yet paradoxically, most amateur astronomers tend to ignore the Moon soon after those first casual observing sessions. Many amateur astronomers regard the Moon as a nuisance because its glare interferes with their ability to observe deep-sky objects like nebulae and galaxies.

It is hard to understand why the Moon is sometimes ignored or even disdained, for it offers at least four distinct advantages to the amateur astronomer which other celestial wonders do not. First, and perhaps most important from a practical viewpoint, the Moon is virtually immune to that bane of the modern amateur astronomer known as light pollution. While most people who adopt amateur astronomy as a hobby dream of glimpsing faint nebulae and galaxies through large-aperture telescopes located under pristine, dark skies unaffected by the lights of our cities and towns, the realities of our modern lives make this dream difficult to attain.

Today, most people have to live in an urban area in order to make a living, and the amateur astronomer wishing to visually observe faint objects must travel great distances to find dark skies, if they can be found at all. And having to travel to dark skies creates its own problems, many of which are hardly insignificant. Packing and unpacking the car with your telescope and accessories is tiring and takes much time, as does the journey to and from the observing site. Unless you are retired, this means that your observing time will be limited to weekends around the new moon. And what if it's cloudy when that long-awaited weekend arrives?

The Moon, on the other hand, does not require you to leave your home to undertake even the most sophisticated observing program. The Moon when full shines at a very bright magnitude –12.6; only the Sun at magnitude –26.7 is brighter. For this reason there is no problem with observing the Moon from the center of London or New York, which certainly cannot be said for nebulae and galaxies. Not having to leave your home means that more observing can be done – even if you have to go to work or school the next day – and you can be ready to observe whenever the skies are clear.

The second major advantage offered by lunar observing is that, unlike deep-sky objects, which are best seen only for a few days around new moon and only during one particular season of the year (depending where in the sky the object is located), the Moon can be observed on almost any night of the year. The inferior planets, Mercury and Venus, can be seen only when their elongations (apparent distance in the sky east or west of the Sun) are great enough, and the superior planets are well seen only around opposition, when the Earth is between the planet and the Sun. For the planet Mars, opposition occurs less than once every two years, and truly favorable oppositions, when the apparent size of Mars as viewed from Earth is large enough to show significant amounts of surface detail to amateur telescopes, occur less than once every decade. In contrast, the Moon, except when new, is virtually always separated from the Sun by a sufficient distance to be seen well, and its angular size, which does not vary dramatically, is always large enough to make observing easy.

The third major advantage of becoming a lunar observer is the incredible wealth of detail that can be observed and imaged, for the rather obvious reason that the Moon is much closer to the Earth than anything else in the night sky (aurorae, meteors, and artificial satellites excepted). At its closest the Earth–Moon distance is a mere 363 263 km (225 721 miles), and this separation never widens to more than 405 547 km (251 995 miles).

Compare these distances to that of the planet Venus, our next-nearest neighbor in space, which even at its closest approach to our planet is still a hundred times farther away. Since the angular size of a celestial object, or any object for that matter, is inversely proportional to its distance, that means that the Moon's apparent size is much larger than anything except the Sun. Roughly one-half degree (1800″) in diameter, the lunar disk is roughly 36 times wider than the next-largest object in the night sky, the planet Jupiter, whose disk at the most favorable oppositions measures a mere 50″ across its equator. To put it another way, the surface area of the Moon visible from Earth is roughly thirteen hundred times the apparent surface area of Jupiter. That much more surface area means that much more can be seen. Many thousands of craters,

mountain ranges, rilles, and other features can be seen with a typical amateur telescope – enough to last several lifetimes.

By contrast, only Jupiter and Mars, among the planets, and only the very biggest and brightest deep-sky objects will show any appreciable detail, and this detail will never approach what can be seen on our satellite. Most galaxies are only faint, fuzzy blurs, even in the largest and most costly amateur telescopes. We must wait until favorable oppositions to glimpse surface features on Jupiter and Mars. Even at opposition, Saturn's disk is largely featureless to the visual observer, save for a dusky belt or two, and it is difficult to make out any detail in its ring system, except for Cassini's Division, which is visible in small telescopes. We cannot see surface features at all on Venus, as they are hidden from our view by that planet's thick atmosphere. Mercury is elusive, and there are few if any surface features with sufficient size and contrast to be seen from Earth. Uranus and Neptune are gas giants whose surfaces were only recently well seen by the Voyager spacecraft. Tiny, distant Pluto looks star-like in even the largest instruments. So the Moon is undisputed champion of all celestial objects when it comes to the visibility of interesting details to observe, draw, photograph, and image.

And if all this were not enough, each of the thousands of lunar features will look different from one night to the next, even over the course of a single night, as the rays of the rising Sun strike the lunar surface at ever-changing angles of illumination. In general, it is best to observe almost any lunar feature when it is close to the terminator (the day/night boundary line that runs north–south across the Moon's disk), that is, when the Sun angle is low, for this condition creates the longest shadows and highest contrast. The only exception to this rule of thumb is for the crater ray systems, which are often best seen at high Sun angles, at times when other surface details become "washed-out." By comparison, deep-sky objects remain quite unchanged from one night (or year!) to the next, except perhaps for variable stars.

The fourth advantage enjoyed by the lunar observer, and maybe the most encouraging, follows as a natural consequence from the other advantages already discussed. Because the Moon is so readily observed and because it is so large and offers such a wealth of easily observed features, the equipment needed to start an exciting and rewarding program of lunar observing is neither complicated nor expensive. Even the smallest telescope, if sturdily mounted, will reward the observer with many hours of enjoyable and fruitful Moon-gazing. And although

CCD cameras and other sophisticated equipment can be a great advantage to the lunar observer (Chapter 12 will tell you how to use these devices to observe the Moon), they are hardly necessary to get started. Now that we've made a convincing case for *why* we should observe the Moon, let's find out *how* to do it.

Choosing a Telescope for Lunar Observing

Other books in this series, most notably *Telescopes and Techniques* by Chris Kitchin and *The Modern Amateur Astronomer* edited by Patrick Moore, have thoroughly discussed the kind of equipment typically used by amateur astronomers, the advantages and disadvantages of each kind, and how to use them. There is no point in repeating all of those discussions here, but I shall mention a few items that beginning lunar observers may find helpful when deciding what kinds of equipment they need to get started.

Although it may seem so obvious as to scarcely require mentioning, a telescope really is indispensable to anyone wishing to study the Moon. Although large (at least 80 mm aperture) binoculars can be used for lunar work if properly mounted (it's almost impossible to hold large binoculars steady enough to observe lunar surface features), you can buy – or if you're so inclined, make – an even larger telescope for about the same amount of money that will show you much more. Of course, that begs the next question – what kind of telescope is best for lunar observing?

In general, any telescope with an unobstructed light path is suitable for observing the Moon. This is because diffused light is one of the great enemies of the lunar and planetary observer. Any obstruction in the light path of a telescope, such as the secondary mirrors in Newtonian or Schmidt–Cassegrain reflectors, will smear out and diffuse the light which it intercepts, thereby reducing contrast, and hence the amount of fine detail, that can be seen. For this reason, refracting telescopes are considered superior to all other optical designs when it comes to observing the Moon's delicate details. Figure 8.1 shows an ideal refracting telescope for lunar observing.

But refractors have two major disadvantages which make them less attractive than reflecting telescopes to amateur astronomers: unless they have a "folded" design, their tubes are very long and cumbersome, requiring massive and awkward mountings, and they

Figure 8.1. An ideal refracting telescope for lunar observing. *Built and photographed by Joe LaVigne.*

are expensive, primarily because their optics consist of two or three objective lenses, requiring the optician to figure four to six separate surfaces which must be compatible with one another.

Most serious lunar observers use another type of telescope – the Newtonian reflector. Figure 8.2 shows my 200 mm (8-inch) Newtonian reflector, which with its long focal ratio is perfect for observing the Moon. The primary mirror at the heart of a

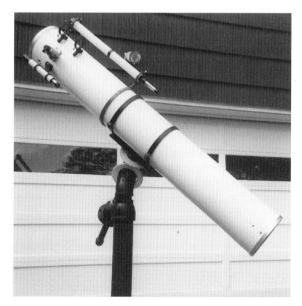

Figure 8.2. The author's 200 mm (8-inch) Newtonian reflector has a long focal ratio, making it a good instrument for observing the Moon. *Photograph by the author.*

Newtonian reflector's optical system requires the optician to figure only one surface, and the mirror's glass can be of a lesser quality than that required for refractor objectives. Newtonian telescopes usually have much shorter tubes than refractors and can therefore take advantage of more compact mounting systems. For all of these reasons reflectors tend to be more affordable than refractors of the same aperture. Another way of looking at this issue is to say that, working with the same budget, the amateur astronomer can buy or make a much bigger reflector than a refractor. As with most things in life, bigger is better. But why is this so?

A telescope's resolving power is directly related to its size: the bigger the diameter of the objective lens or mirror, the more resolving power the observer will enjoy. Resolving power can be calculated using the Dawes limit, named for the British amateur astronomer William Dawes:

$$R = 110/D,$$

where R is the resolving power in seconds of arc, and D is the diameter of the telescope's lens or mirror in millimeters.

Thus, a 250 mm (10-inch) telescope like the author's (see Fig. 8.3) has a Dawes limit of 0.456″, which means that an observer using that telescope could expect to split a double star whose components were separated by that distance, under ideal seeing conditions. But does this calculation mean that an observer using that same 250 mm telescope under similar conditions would be able to see features on the Moon whose apparent size is 0.456″ across? Well, yes and no.

Because light from linear features tend to survive its passage through the turbulence of our planet's atmosphere better than for two-dimensional features, the same telescope will show lunar rilles, which are essentially lines, that are much finer than the smallest visible craters, craterlets or mountains, which are all essentially two-dimensional objects. Table 8.1 shows the smallest details visible on the Moon through telescopes of different apertures.

At least two conclusions can be drawn from Table 8.1. First, it is clear that bigger apertures allow you to see finer details on the Moon. So is it best to acquire a reflecting telescope of the greatest aperture possible, given ones personal budget? Not necessarily, for Table 8.1 also shows that very small lunar features are visible even in modestly sized amateur instruments. Most of the classic maps of the Moon showing the vast majority of detail visible through amateur telescopes were made by observers using small telescopes. Wilhelm Beer and Johann Heinrich Mädler, for example, used a telescope of

Figure 8.3. The author's 250 mm (10-inch) Schmidt–Cassegrain telescope. *Photograph by the author.*

Table 8.1. The smallest lunar details that can be seen through telescopes of different apertures

Aperture (mm)	Smallest crater (km)	Smallest rille (Detectable, m)	Smallest rille (Fully resolved, km)
25	14.5	805	6.44
50	7.25	403	3.22
76	4.83	268	2.14
102	3.62	201	1.61
152	2.40	137	1.08
203	1.76	101	0.81
254	1.45	82	0.64
305	1.21	64	0.53
381	0.97	55	0.40

only 95 mm ($3\frac{3}{4}$ inches) in constructing their great map of the Moon! So the beginning lunar observer might do well to acquire a small refractor of high quality – especially if the alternative is a reflector of dubious quality.

I would offer one last word of advice in this regard, based upon thirty years of experience in observing the Moon: the obsession with detecting the tiniest lunar features possible is greatly overrated. Common sense should tell us that the most beautiful and interesting lunar features are unlikely to be the smallest ones. While it may be interesting to see how many craterlets you can detect on the floor of Plato with your telescope, if you've seen one craterlet, you've seen them all! Beyond a certain point, the observer using larger apertures is rewarded only with greater numbers of very small and undistinguished lunar features, none of which will ever be as exciting as the larger, easier-to-see features. Moreover, seeing conditions rarely allow the resolving power of a telescope to be utilized to maximum advantage – more on this later in the chapter.

For these reasons, my opinion is that 410 mm (16 inches) is probably the largest aperture that an amateur astronomer will find practical to use to observe the Moon, under normal circumstances. If one is lucky enough to use larger apertures, and one is a very experienced observer, it may be possible to wring out very fine lunar details that elude smaller instruments. Patrick Moore and Hugh Percy Wilkins in the 1950s used the great 0.84 m (33-inch) Meudon refractor to make ground-breaking visual observations of several intriguing lunar features such as the "Cobra Head" of Schröter's Valley, and more recently the author did the same with the 40-inch (1 m) refractor at Yerkes Observatory, but few amateurs will ever have access to such large and historic instruments. For the beginning lunar observer, however, a 75 mm (3-inch) refractor or a 150 mm (6-inch) reflector will reveal a bewildering amount of lunar detail, and will be more than sufficient as a first telescope.

If the fledgling lunar observer decides upon a reflecting telescope, there are two considerations to keep in mind. For one, consider only reflectors with focal ratios of $f/6$ or longer, preferably $f/8$. My 250 mm (10-inch) telescope, which is a Schmidt–Cassegrain variety of reflector, has a focal ratio of $f/13.5$! Focal ratio (sometimes abbreviated f-ratio) is simply the focal length of the telescope divided by the aperture. Thus, a 250 mm (10-inch) reflector with a focal length of 2 m (80 inches) has an f-ratio of 8, written as $f/8$. Focal ratios faster (less than) $f/6$ tend to have too many aberrations to be fit for lunar observing. It is extremely difficult (and expensive) to make a mirror faster than $f/6$ with the exacting specifications required for lunar observing. Telescopes with mirrors faster than $f/6$ are primarily designed for deep-sky observing, where the optical aberrations commonly found in these fast mirrors do not appreciably degrade the image of diffuse objects like nebulae and galaxies.

Something else to keep in mind when buying or making a reflecting telescope that will be used to observe the Moon is that it is highly desirable, even imperative, to make sure the secondary mirror is as small as possible. This consideration goes hand in hand with the first: long f-ratio instruments will allow the use of much smaller secondary mirrors than telescopes with fast primary mirrors. Low-profile focusers, which are now very common on commercially made telescopes, also help to reduce the size of the secondary mirror that must be used. This is because both long focal ratios and low profile focusers act to decrease the size of the light cone which the secondary mirror must intercept and reflect into the eyepiece.

Why do we want the secondary mirror to be as small as possible? To minimize the effects of diffraction, which we have briefly discussed in this chapter. The smaller the secondary mirror of your reflector, the more it will perform like a refractor of similar aperture. The less diffraction your telescope has, the more you will see when you point your telescope at the Moon.

The major drawback of Schmidt–Cassegrain telescopes (SCTs), which have become immensely popular in the last twenty years, is that they suffer from lots of diffraction. This is because their secondary mirrors are, by necessity, set very close to the primary mirror and are therefore very large, creating a large central obstruction. Ideally the central obstruction should be less than 20% of the aperture. Many SCTs have central obstructions of 35% or more. My SCT has a central obstruction of 30%. Moreover, the Schmidt corrector plate, which is the thin lens at the front end of an SCT, easily accumulates dust and dew, scattering even more light.

Besides telescope design, the following factors will go a long way to determining how useful your telescope will be in making observations of the Moon:

1. The quality of the optics.
2. How well the optics are collimated.
3. The local seeing conditions.
4. The skill of the observer.

Whatever telescope design you choose, the first of these factors is more important than any other in determining whether you've made the right choice: how good are the optics? A well-collimated reflector with a well-made long f-ratio mirror will easily outperform a poorly made or poorly collimated refractor. Many "fast" mirrors that are found in popular Dobsonian-style reflectors are ill-suited to lunar observing, as their optical surfaces are not figured accurately enough to give the contrast and sharpness necessary to bring out the fine detail in craters, rilles, and mountain peaks on the Moon.

Unfortunately, the same defects are too often found even in the mirrors of the more expensive SCTs that are also very popular. My SCT was made in the 1960s, and optical tests suggest that its mirror is much better than the typical mirrors (of which I have tested dozens) found in today's SCTs.

In some ways this is a "Catch 22." If you are a beginning observer who wishes to start observing the Moon, how will you know if your telescope has a good mirror or lens before you buy it? This is not an easy question to answer, but there are at least three steps you can take to maximize your chances of getting good optics. First, ask more experienced amateur astronomers in your area to recommend a telescope or optical company before you spend your money. If possible, join your nearest amateur astronomy club or society and get to know its members. They will almost certainly be happy to let you look through their telescopes. If you are thinking of buying a particular brand of telescope, you may be able to "test-drive" that instrument if someone in your club already has one.

Second, try to find out whether the telescope's maker guarantees that the optics are *diffraction limited*. This *should* mean that the optics will resolve objects to the Dawes limit described earlier in this chapter, under perfect seeing conditions, but in practice telescope-makers use the term "diffraction limited" to mean different things. Still, the poorer optics are not usually advertised as "diffraction limited." Third, make sure that the telescope has some sort of money-back guarantee. Then, as soon as your telescope arrives, have more experienced members of the local astronomy club check it out for you. They should be able to star-test the telescope and perform other in-the-field tests that will determine if it is performing up to par.

It must be emphasized that no telescope, no matter how good the optics are, will perform adequately if those optics are not properly aligned or collimated. Collimating different types of telescopes is covered in great detail in another book in this series, *Telescopes and Techniques: An Introduction to Practical Astronomy*, by Chris Kitchin. If you buy a new telescope, you should assume that it will arrive un-collimated. The very first thing you should do after unpacking and assembling the telescope is to re-collimate it. Even if it was collimated at the factory, it will almost surely require re-adjustment, as it will have been jostled during shipping.

It is also worth mentioning that for any kind of lunar observing, which will typically involve high magnifications, your telescope, whatever kind it is, must be solidly mounted, and preferably on an equatorial mount, the kind that allows you to com-

Figure 8.4.   A drive corrector, which allows for the Moon's 13°/day easterly drift. *Photograph by the author.*

pensate for Earth's rotation to keep the object you're observing in the field of view for long periods of time. Since the Moon is by far the most highly detailed astronomical object you can observe, you will find yourself spending long periods of time at the eyepiece trying to pick out these fine details, and you will get frustrated if the features you are studying continually drift from the eyepiece field.

It therefore helps to equip the telescope's equatorial mount with an accurate clock drive, a device that rotates the telescope around the polar axis (usually a shaft that is part of the equatorial mounting, aligned with the center of the Earth's rotation at a place in the sky near the north pole star). However, if you really want to keep the Moon from drifting away you will also need a "drive corrector" to adjust for the fact that the Moon moves against the "fixed" background stars. A drive corrector like the one shown in Fig. 8.4 corrects the clock drive to allow for the Moon's 13° daily drift eastward across the sky.

Local seeing conditions will probably do more than anything else to affect the performance of your telescope – if you have good, well-collimated optics, that is. If you can observe from a grassy area instead of a concrete or asphalt-covered driveway, by all means do so. Seeing is discussed more thoroughly a little later in this chapter.

Finding Your Way Around the Moon

When it comes to the skill of the observer, there is no substitute for the two "p-words": practice and patience. It is a well-known fact that the neophyte observer commonly misses details that are obvious to the more experienced observer. Practicing your observing technique over and over again at the eyepiece is guaranteed to reward you with the ability to see many things that were invisible to you when you began your observing career. But you must have patience, something that is fast becoming a rare quality in our modern, harried world. Unlike watching TV, you cannot "channel-surf" when doing visual astronomy, especially on complex objects like the Moon.

This has to be emphasized – you must develop the patience to focus on the feature you are interested in for long periods of time, for the longer you look, the more detail you will notice. Typically, making a detailed drawing of a lunar feature will take the experienced observer about half an hour. Making drawings of lunar features is the best way of learning about them – the very exercise of drawing a crater, for example, forces you to take notice of details like the size and shape of the crater's walls, how steeply the walls slope, whether the crater has a central mountain peak (or peaks), and whether the floor of the crater is smooth, pocked with craterlets, or perhaps scarred by one or more rilles. But before you can begin drawing the Moon's features, you must learn your way around our satellite.

The very first exercise for the new lunar observer should be a simple visual survey of its changing face through a one-month lunation, getting to know the major surface features, learning its basic geography. With thousands of craters and dozens of major mountain ranges, this is no easy task! But do not despair – you don't have to memorize all that you see, just get acquainted with it. The first step is to find out when new moon occurs. Almost any good calendar will show this. The *BAA Handbook* and the Royal Astronomical Society of Canada's *Observer's Handbook* also give this information. But knowing the phases of the Moon will not tell you right off which craters and mountains will be visible near the terminator on any given night. The BAA Lunar Section's Website has a list of lunar features near the terminator for each night of the current month-make sure you look at this listing before you go out to observe.

But a list of features to observe, while useful, still won't help you locate the features on the Moon as viewed through your telescope. To do that you will need a good lunar atlas. There are many beautiful lunar atlases that are very artfully drawn, but most of these are hard to use at the telescope. For many years, my favorite atlas for this purpose has been the *Photographic Lunar Atlas* compiled by Commander

Henry Hatfield. Commander Hatfield, who served in the Royal Navy as a navigator and hydrographic surveyor, is a past director of the BAA Lunar Section. The Hatfield atlas is superb because it accurately represents the Moon as it is seen through amateur instruments, and is also treasured for its clear and easy-to-use outline maps of the Moon's major features.

Figure 8.5 shows a portion of one of the line drawing maps from the Hatfield atlas. Amazingly, the work for this atlas was carried out using a home-built reflector and ordinary photographic equipment – that should be an inspiration to any aspiring amateur astronomer wishing to make a contribution to this field. Originally published in 1968 and for many years out of print, the Hatfield atlas has been updated and reprinted by the publisher of this book. For the more advanced observer, the Lunar Quadrant Maps (see Fig. 8.6), drawn by the staff of the University of Arizona's Lunar and Planetary Laboratory, are highly recommended.

Eyepieces and Filters

There are two different ways you can go about learning how to find and identify lunar features. You can simply aim your telescope at a spot on the Moon and try to identify the features you see in the eyepiece by referring to your lunar atlas; or you can do it the other way round, by picking objects listed on the BAA Lunar Section's Website, then using your lunar atlas to aim your telescope at the appropriate places on the Moon's surface. Both techniques have their own merits, and you should probably alternate between the two. In either case, it is recommended that you start by using the lowest-power eyepiece possible, since it is always easier to work with a wider field of view, giving you the biggest possible "picture" of the lunar surface. Once you've made a positive identification of an object in the eyepiece at low power, you can switch to higher-power eyepieces to see the object in greater detail. With practice, you will become quite adept at identifying new objects that appear in your eyepiece or tracking down objects on your list.

So you should start with at least two eyepieces in your collection – a low-power eyepiece that allows you to fit all or most of the Moon into the field of view, and a higher-power eyepiece, of at least twice the magnification of the low-power one. The magnification of an eyepiece depends on the focal length of the telescope it is used with – magnification is just the telescope's focal length (typically measured in

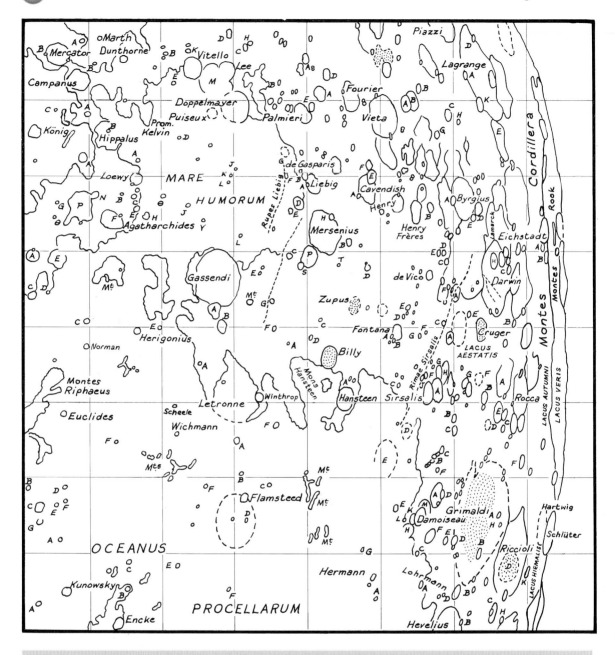

Figure 8.5.   One of the outline maps from the Hatfield atlas. *Copyright © British Astronomical Association.*

millimeters) divided by the focal length of the eyepiece. Thus, a 25 mm eyepiece when used with a telescope having a focal length of 2.5 m (98 inches) will yield a magnification of ×100.

I typically use a 26 mm eyepiece in my 250 mm (10-inch) telescope, which has a focal length of 3.429 m (135 inches). This gives a magnification of ×132. This is my "low-power" eyepiece, but ×132 is considered a fairly high power for most telescopes

and observers. However, my experience allows me to aim the telescope accurately at the object I want to observe, even at this magnification. Most beginning lunar observers will probably want a much lower-power eyepiece – something that yields a magnification of ×50 or ×60. For high-power views of lunar features, I prefer a 10 mm eyepiece, which gives me a magnification of ×343. This is very high power for most observers, and the beginning lunar

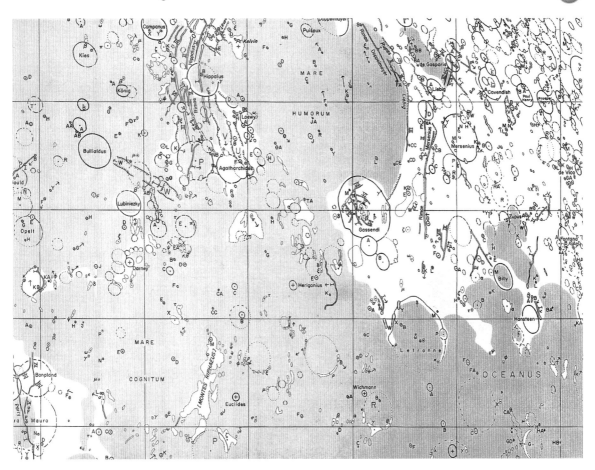

Figure 8.6. A section of one of the University of Arizona's Lunar and Planetary Laboratory's *Lunar Quadrant Maps*. *University of Arizona's Lunar and Planetary Laboratory.*

observer will probably be more comfortable using an eyepiece that gives a magnification no higher than ×150 to ×200. It should be mentioned that our old friend *seeing* comes into play here, too. When the seeing is very steady, much higher magnifications can be used than when the air is unsteady. When seeing is poor, you will be restricted to using your lower-power eyepieces.

I am often asked which designs of eyepiece are best for lunar observing. This is a tough question to answer, since there is a bewildering array of eyepiece designs on today's market. Figure 8.7 shows several excellent eyepieces for lunar observing. There are four main criteria by which eyepieces may be judged:

1. How well they correct the common aberrations.
2. The degree of contrast shown on objects in the eyepiece.
3. The eye relief.
4. The field of view.

Figure 8.7. Several excellent eyepieces, ancient and modern, for lunar observing. *Photograph by the author.*

A well-corrected eyepiece will be free from the various types of *aberration*: spherical and chromatic aberration, astigmatism and coma. An in-depth discussion of these aberrations is beyond the scope of this book. Suffice it to say that stars should appear as pinpoints across the field of view, with no false color, and not distorted into non-pinpoint shapes. If stars look sharp and "pinpointy" in an eyepiece, the chances are that the Moon will look good in it too. In general, the better-corrected eyepieces contain more lenses (or "elements") – up to nine – and are more expensive the more lenses they contain.

The *contrast* achievable with an eyepiece will depend on its optical design, the quality of the glass and the coatings, and, often overlooked, the degree of baffling inside the eyepiece to absorb stray light reflections. (Baffling is achieved by the use of anti-reflective coatings on the optics, special paints on the metal surfaces that are highly effective at absorbing light, and carefully designed "field stops" placed in the optical path.)

Eye relief is simply how far back from the eye lens of the eyepiece you can position your eye when you are observing. Poor eye relief means having to hold your eye right up next to the eye lens, which is tiring and difficult for any observer to do. It is harder to spend much time looking for subtle details on the Moon if your eyepiece has poor eye relief. Better eyepieces offer eye relief of up to 20 mm – meaning that you can hold your eye almost an inch away from the eyepiece and still focus comfortably. Good eye relief is a must if you need to wear glasses to observe, but even if you don't, they make observing sessions infinitely more comfortable and enjoyable.

Field of view is the width of the eyepiece field. Having a wide field of view is particularly desirable when observing extended objects like star clusters or nebulae, but it is a little less critical when observing lunar features, since these usually fit with ease into the smallest fields of view. In fact, it is sometimes desirable to have a smaller field of view when observing lunar detail, since surrounding objects that might otherwise be a distraction are blocked out. Eyepieces with good eye relief tend to have smaller fields of view, and vice versa. If you have to choose between eye relief and large field of view when selecting an eyepiece for lunar observing, go with the greater eye relief, as this factor is much more important for lunar work.

In general, the more expensive eyepieces, by makers like Clavé, Pentax, Takahashi, TeleVue, Vixen, and Zeiss, will be better corrected, produce greater contrast, and, depending on their design, offer either maximum eye relief or maximum field of view. Eyepieces made from glass that incorporates the rare earth element lanthanum have become quite popular in recent years, as they offer good contrast and superb eye relief. The quality of the eyepiece is more critical for higher powers, so if you buy a bargain eyepiece, make sure it is used only for low-power observing.

As for accessories, basic lunar visual observing requires very few. One accessory I regard as indispensable is a neutral density filter. This is simply a polarized, anti-glare filter that can be bought at your local camera shop. I use a "variable" neutral density filter especially adapted for use with for 32 mm (1.25-inch) diameter eyepieces. It can be adjusted to compensate for the Moon's brightness, which in turn depends on its phase (full moon being brightest) and whether the Moon is dulled by being low in the sky (by atmospheric absorption), or by clouds or smog. The Moon is such a bright object that it can actually be uncomfortable to observe in a standard telescope unless some kind of anti-glare filter is used. Employing such a filter will allow your eye to relax, and will often improve the contrast of lunar features.

Some lunar observers find color filters useful. Medium Blue (Wratten #80A) is considered to be best for enhancing lunar surface detail, but Light Yellow (#8), Yellow-Green (#11), Yellow (#12), Deep Yellow (#15), Orange (#21), Light Red (#23A), Light Green (#56), Green (#58), and Pale Blue (#82A) are also considered effective, to varying degrees, by experienced observers. Color filters tend to be affordable, so most observers have at least a couple of them in their eyepiece box.

Of course, clouds and bad weather will keep you from observing the Moon every night during your "get acquainted tour," but do the best you can. You will at first be surprised by how quickly the terminator sweeps over the Moon's surface from one night to the next – a crater that was favorably seen with sharp contrast on the terminator one night will be largely washed out the very next night, and perhaps barely noticeable the night after that. You will probably discover certain features on the Moon that will become your favorites, and you will want to observe these in greater detail during the next lunation. Once you've had some practice finding specific features on the Moon, you will be ready to begin drawing them.

Chapter 9

Drawing Lunar Features

Why go to the trouble of drawing lunar features? First, the exercise will force you to sharpen your visual observing skills. Later in this chapter there is a checklist that will help you to better describe what you are seeing. This exercise will also give you a chance to apply what we've learned about lunar geology. Second, drawings are the best way of preserving your observations – a picture is worth a thousand words, as they say. This will prove useful when you want to remember what you've already observed, and to compare observations of the same features at different Sun angles or at different stages of a lunation.

Second, the appearance of a particular feature will be different before the Moon is full than after the Moon is full, even when it is on the terminator on both occasions. This is because in the first case the illumination will come from the east side of the Moon, whereas in the second case it will come from the west side. Third, the human eye is adept at detecting subtle features that are difficult, if not impossible, to capture on film or CCD. If you decide to try your hand at lunar photography, you will quickly discover that even your best photographs show only a fraction of the detail that your trained eye can make out.

Getting Started

Drawing lunar features requires some preparation. You will need the following supplies:

1. A good *source of light*, so you can see what you're drawing. Since the Moon is such a bright object there's no need for the precaution of using a red filter over the light source that is necessary when

observing other celestial objects. If you are observing from near your home you can probably use an ordinary lamp. The light source should be free-standing, to leave your hands free to adjust the telescope and make the drawing. Your observing form or drawing paper should be secured to a clipboard, so that it doesn't move around while you are drawing. Some lunar observers prefer a simple piece of plywood or other hardwood, to which they clip or tape the observing form.

2. *Observing forms*, like the one in Fig. 9.1. The observing form will help you to draw the lunar object at an appropriate scale, and it will prompt you to include all of the necessary data that will make your drawing scientifically useful. Some accomplished lunar artists prefer to use blank drawing paper, on which they later fill in the necessary data.

3. A variety of *pencils* of varying hardness. Your local art supplies shop should be able to recommend a good selection of pencils if you tell them what you wish to draw. You may want to make the outline drawing of your lunar feature with a harder pencil. The soft pencils are used for shading lunar relief. One of the pencils should be an off-white colored pencil, for drawing very bright parts of lunar features. Some lunar observers also like to use a black colored pencil, for drawing deep shadows, while others prefer gray colored pencils of varying shades. You can also use charcoal, but this usually requires greater skill from the artist if it is to be used effectively.

4. *Artist's stumps*, which are tightly rolled wads of paper that look somewhat like short pencils, for smearing soft pencil shading. Once you have

Observer's Form-Lunar Topographic Features

Observer: _____ . Address: _____ .

City/Town: _____ . Country/Postal Code: _____ .

Telephone _____ . E-Mail: _____ .

Telescope: _____ . Focal Ratio: _____ .

Magnification: _____ . Filter(s), if any: _____ .

Name of Lunar Feature (s): _____ .

Date Observed: _____ . Selenog. Co-longitude: _____ .

Obs. Begins (U.T.): _____ . Obs. Ends (U.T.): _____ .

Seeing/Scale: _____ . Transparency: _____ .

Drawing:

Notes:

Figure 9.1. A sample lunar observing form. *Prepared by the author.*

learnt how to use a stump you will be able to achieve very subtle differences in shading, and also to obtain a uniform degree of shading over a wide area, for example across the floor of a crater.

5. An accurate *timepiece*, for recording Universal Time when you start and stop your observation. This piece of equipment is essential no matter what kind of astronomical observation you are making.

Once you have the necessary supplies, you should select one or two features to draw in one observing session. Large craters make a good start, as they contain most of the different types of topography you are likely to encounter in learning to draw the Moon. I highly recommended that before you begin trying to draw a crater and the details inside it, you start with an accurate prepared outline of the crater. This might sound like cheating, but it's not! In practice, it is extremely difficult for even the most experienced lunar observer to portray the size and shape of a crater's outline accurately. The proof of this is seen in the outlines of the same crater as drawn by different classic lunar mappers – no two renderings of the same feature were ever the same! It's not just OK to use pre-drawn outlines for lunar craters, it's smart, too. If the basic size and shape of the feature are not accurate, the rest of the drawing will be of dubious value.

The best way of getting an accurate, pre-drawn crater outline is to trace it from a photograph. But a word of warning is in order here: make sure that the photograph you are tracing from was taken through an Earth-based telescope. Only photographs taken from the Earth will accurately portray the relative sizes and shapes of lunar features as seen through the eyepiece. In the 1960s, the United States and the Soviet Union sent many spacecraft to the Moon for the purpose of mapping its surface in unprecedented detail. Foremost among these probes were the Lunar Orbiters, launched by the United States in 1966–7, which transmitted back to the Earth hundreds of photographs of the Moon's surface at resolutions of 500 and 65 m (1640 and 215 ft).

The results of the Orbiter surveys were published in several books, including the *Lunar Orbiter Photographic Atlas of the Moon* by Bowker and Hughes. While the Orbiter atlases are extremely useful for verifying very fine detail at the limit of the observer's equipment, they are not suitable for tracing the outlines of lunar features before drawing them at the telescope. Lunar Orbiter photographs were made from angles that can never be duplicated by Earth-bound observers, and make the shapes of lunar features look very different than the often foreshortened views from Earth. When using the Orbiter photos to verify features that are at your observing limit, don't look at them *first*, otherwise you may convince yourself you are seeing details that you really can't! Look at them afterwards to check the authenticity of your visual observations.

The Hatfield lunar atlas is especially suitable for preparing outlines of lunar features, since it includes not only Earth-based photographs but also outline charts of the major craters and other features – the tracings have already been done for you. Most of the tracings are on too small a scale, however, to be used directly, and you will probably want to enlarge them on a photocopier. Figure 9.2 shows an outline of the lunar formation Clavius prepared from the Hatfield atlas. Once you have your crater outline, the next step is to hope that the night sky will be clear so that you can zero in on your target. If luck is with you, you will be ready to make your first lunar drawing.

Before you begin to draw, you should come to grips with the fact that drawing lunar features is one of the most challenging exercises you can undertake as an amateur astronomer. Unless you have artistic training you will find it difficult to draw what you are seeing through the eyepiece. Unlike deep-sky objects, which are essentially two-dimensional in even the largest telescopes, the features of the Moon appear obviously three-dimensional, even in small telescopes.

What's more, most deep-sky objects show very little variation in brightness or shading across their width: a galaxy might have a central core that is brighter than its spiral arms, while a nebula might

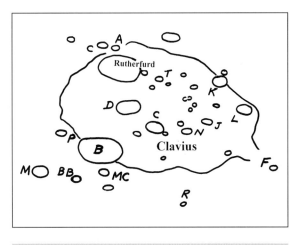

Figure 9.2. An outline of the lunar formation Clavius, prepared from the Hatfield atlas.

have certain parts that are a little brighter than others. But even the simplest lunar feature displays an astonishing range of bright and dark areas. And last, but certainly not least, unlike a still-life object that patiently sits in one place, waiting for you to draw it accurately, the Moon's terminator waits for no observer! You must record all you see in half an hour or less – any longer, and the scene through your eyepiece will have changed too drastically from the time you began your observation.

For these reasons, it is probably not a bad idea to enroll in a basic sketching class at your local college to get some fundamental training in sketching techniques that will prove useful at the eyepiece. Tell the instructor why you are there and what you wish to draw – they may be able to offer expert advice that will be specific to drawing lunar craters and mountains, which, after all, is not much different than drawing a terrestrial landscape, albeit from a distance of several hundred thousand kilometers!

Questions to Ask

The first step in making a lunar drawing involves no drawing at all. Before you draw anything, whether it be the proverbial bowl of fruit or a crater on the Moon, you should study your subject carefully, trying to "understand" it. You must ask yourself just what it is you are looking at, and which of its characteristics you wish to represent in the drawing. Once you have thought about these things, you can consider how to approach the task. Before attempting to draw a lunar feature you should ask yourself a series of questions that force you to identify the feature's most salient aspects, jotting down the answers for your own reference. The following questions are appropriate for a lunar crater, or walled plain:

1. What are the relative size and shape of the crater's outline? What geometric shape or shapes does it resemble? Even if you are starting with a pre-drawn crater outline instead of trying to draw the it freehand, it is useful to go through this exercise.

2. What do the crater's walls look like? How gently or steeply do they slope? Do some parts of the walls slope differently from others (they almost always do)? Make careful notes about the slopes of the crater walls.

3. Are the walls broken or heavily eroded in places? What are the shapes and lengths of the shadows cast by the walls, and how are these shadows affected by breaks or erosions in the walls?

4. Are the interior of the walls terraced, or smooth? Remember that terracing is often a subtle feature – you may notice terracing after you have carefully observed the crater for awhile, even if you initially didn't see it.

5. Does the crater have one or more central mountain peaks or hills? What is the shape of their base? What shapes are the shadows they cast?

6. What are the relative positions, sizes, and shapes of craterlets within the crater? Are these craterlets arranged in groups or chains? Almost all craters or walled plains have at least several craterlets.

7. Does the floor of the crater contain less depressed areas that look different from craterlets? What are their relative positions, sizes, and shapes?

8. Does the crater floor contain any rilles or clefts? How long and wide are they, and in which directions do they run? Do they intersect or contain any craterlets, or appear to emanate from them?

9. Is one side of the crater floor higher than the other? Are there any obvious fault or scarp lines visible? Does the floor contain any lunar domes?

10. Is the crater floor a uniform shade of gray, or does it appear to vary in color?

This list of queries is not meant to be exhaustive, and creative thinkers will no doubt come up with others. If you are drawing features like mountain chains or rille systems on the maria you will want to ask slightly different questions, but craters and walled plains are the most popular targets for amateur lunar astronomers. The value of going through such an exercise is illustrated in Fig. 9.3, a drawing of the lunar formation Cleomedes made by Colin Ebdon. Figure 9.3 has been annotated in line with the ten questions just posed, and it shows how asking the right questions can help to fix in your mind the details of the feature you want to draw.

After you've answered these questions and familiarized yourself with your subject, the next step is to make simple outlines of the positions and shapes of the various topographic features of your crater. You may wish to start with the central peak. Don't worry about accurately representing its actual appearance at first – just draw the outline of its base, trying to be as precise as possible in its location and size. You will probably discover that, like most beginning lunar artists, you tend at first to overstate the size of objects within the crater's boundaries. You will improve at getting the size of central peaks and craterlets right with time.

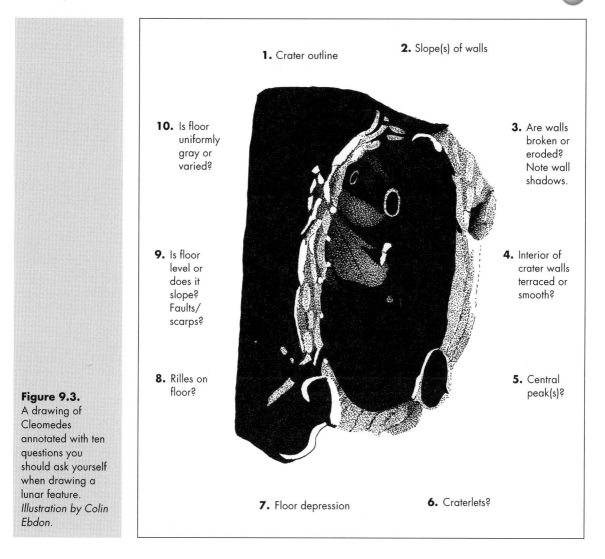

1. Crater outline

2. Slope(s) of walls

10. Is floor uniformly gray or varied?

3. Are walls broken or eroded? Note wall shadows.

9. Is floor level or does it slope? Faults/scarps?

4. Interior of crater walls terraced or smooth?

8. Rilles on floor?

5. Central peak(s)?

7. Floor depression

6. Craterlets?

Figure 9.3.
A drawing of Cleomedes annotated with ten questions you should ask yourself when drawing a lunar feature. *Illustration by Colin Ebdon.*

Adding craterlets is the next logical step of your drawing – again, just make draw the outlines of these craterlets, and don't worry about their shadows or internal shadings. After you have accurately located the central peak(s) and any craterlets, you should find it easier to draw in any rilles or clefts running across the crater floor. Figure 9.4 is a drawing of the lunar formations Pitatus and Hesiodus which shows what your drawing will look like when you have got this far. Once you have outlines of all the topographic features drawn in, you are ready to start shading the outlines to give them three dimensions, and to indicate light and dark.

Many lunar observers are content to stop here with their drawings. There is already substantial value in an accurate line drawing of a lunar feature, and most journals and newsletters devoted to lunar observing will be happy to receive such drawings. A line drawing may not be as artistically beautiful as a full shaded sketch, but it can contain almost as much information. If you are interested in documenting how many craterlets you were able to see on the floor of Plato, for example, a line drawing is more than sufficient. You will be able to compare such a drawing to a high-resolution photograph of the same feature to see how well your telescope (and your eyes!) performed.

Of course, a fully shaded sketch does have its advantages. Not only will it much more realistically portray what you saw in the eyepiece, it will allow you to include information that a line drawing cannot. For example, a skillful sketch will accurately convey the shape and slopes of crater walls, whereas a mere line drawing will not. In general, you will want to start shading your drawing by concentrating on the shadows cast by the crater walls and central peaks. Again, it is sufficient to start by simply drawing the outlines of these shadows, which

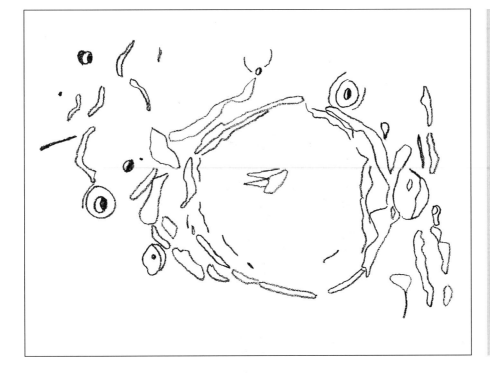

Figure 9.4. A preliminary drawing of Pitatus and Hesiodus. *Illustration by the author.*

usually have jagged triangular shapes to them. The closer a crater lies to the terminator when you draw it, the longer and more jagged these shadows will appear. You should try to shade in the shadow outlines to duplicate the "blackness" of each shadow. Not all shadows are equally dark. If you don't have time to color in the shadows, try to make notes about which shadows are darker and which are lighter.

The opposite of a shadow is a bright area, so it makes sense next to highlight, perhaps with an off-white colored pencil, any part of the crater wall, central peak, or a floor feature that appears significantly brighter than the uniform gray background of the crater floor. Usually, the side of the crater wall facing the direction sunlight is coming from will appear bright, while the other side of the same wall will cast shadows. The same is true for the central peak, if any, and for the interiors of craterlets. One half of the craterlet's interior will be brighter than the lunar background, while the other half will be blacker.

Once you have drawn in the principal shadows and bright areas, you can finish your drawing indoors. This is a good time to color in the lunar background, which is a uniform gray. Many lunar observers prefer to sprinkle powdered graphite – which you can scrape from a #2 (HB)pencil – over the entire drawing, smoothing it over with an artist's stump or a cotton ball. Make sure the background is

indeed uniform, and take care not to make it too dark. Take care also not to fill in any bright areas with the graphite, since you want these to stay off-white. Figure 9.5 is the finished drawing of Pitatus and Hesiodus. Do not be discouraged if your early efforts don't look this good – practice makes perfect, and there's no substitute for artistic talent. Nevertheless, even finished drawings that aren't this pretty will be a valuable record of what you've observed.

You can practice each of the lunar drawing techniques described above by simply copying photographs of lunar features, like those that appear in the Hatfield atlas. If you can accurately copy a photograph, chances are you will be able to draw a lunar feature as it appears in the eyepiece. Copying photographs will also teach you what different kinds of lunar features look like, and how to portray them on paper. As with tracing crater outlines, you should stick to photographs made with Earth-based telescopes, as these will more truly represent the shadows and shading that you will see in the eyepiece.

Some Important Data

When you have finished your drawing, what information should you add to it to make it scientifically useful? The sample drawing in Fig. 9.1 includes all of

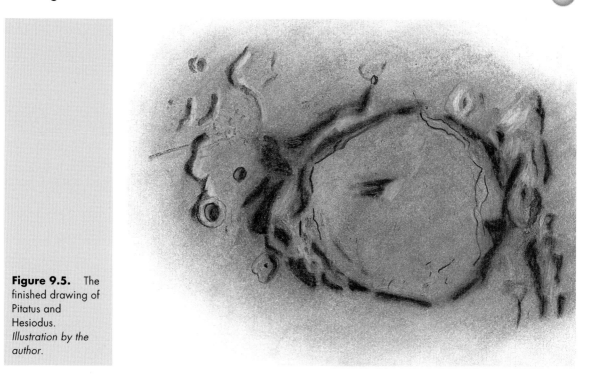

Figure 9.5. The finished drawing of Pitatus and Hesiodus. *Illustration by the author.*

the relevant information that you need to record. Some of this information is rather obvious and straightforward: you should make a note of the date and times of observation, the type of telescope you are using, and the magnification employed. Notice that the times of observation are the times when you started and finished the observation, recorded in Universal Time (UT). Typically a detailed lunar drawing will take about thirty minutes of time at the telescope. It is difficult to record all the details in less time than that, and if you take much longer than half an hour, your drawing will lose accuracy because the rapidly changing Sun angle will change the appearance of the feature you are drawing.

Some of the data that must accompany your drawing may seem rather esoteric to the new observer. They includes the mean selenographic co-longitude, selenographic latitude, and measures of seeing and transparency. What is all this? Let's start with selenographic latitude and longitude. This is just the latitude and longitude (*x* and *y* positions) of an object on the Moon's surface, similar to terrestrial latitude and longitude for a point on the surface of the Earth. You can read off the selenographic latitude of the lunar feature you are drawing from any good lunar map, or you can look this information up in a catalog of lunar features.

Record the selenographic latitude, but not the longitude. Instead, for the latter measure you will record the selenographic *co*-longitude, which is the lunar longitude of the sunrise terminator. Co-longitude is more nearly accurate than phase for indicating the position of the Sun relative to the Moon because it has been corrected for the Moon's librations. The classic observing guides to the Moon often contained a chapter or section that described in painstaking detail how to calculate selenographic co-longitude. Unless you enjoy arithmetic gymnastics, calculating co-longitude for yourself is not recommended. Nor is it necessary, for you can look up this quantity in the *BAA Handbook* for any given date, at midnight. Figure 9.6 is a sample page from the *BAA Handbook* showing computed selenographic co-longitudes for the year 2000. For times other than midnight, you can interpolate. For example, if you are observing on June 30, 2000 at 11.00 p.m. from Los Angeles, California (which, taking into account Daylight Savings Time, is 06:00 UT for July 1, the following date) your co-longitude will be about one-fourth of the way between the values for July 1 and July 2 – that is, one-fourth of the difference between 260°.3 and 272°.6, or about 263°.4. If you don't wish to go through this exercise, the BAA Lunar Section's Website allows you to make detailed co-longitude calculations. Make your computer do the work for you!

The scales of transparency and seeing require a little explanation, and their use requires some practice on your part. Simply put, transparency means the clarity of the sky, and is usually measured by

SUN'S SELENOGRAPHIC COLONGITUDE

Day	Jan.	Feb.	Mar.	Apr.	May	June	July	Aug.	Sept.	Oct.	Nov.	Dec.
	°	°	°	°	°	°	°	°	°	°	°	°
1	201·7	218·7	211·5	229·1	235·1	253·7	260·3	279·3	298·0	304·0	321·8	326·8
2	213·9	230·8	223·7	241·3	247·3	265·9	272·6	291·5	310·2	316·2	334·0	339·0
3	226·1	243·0	235·9	253·6	259·5	278·2	284·8	303·8	322·4	328·4	346·1	351·2
4	238·2	255·2	248·1	265·8	271·8	290·4	297·1	316·0	334·7	340·7	358·3	3·3
5	250·4	267·4	260·3	278·0	284·0	302·7	309·3	328·2	346·9	352·8	10·5	15·5
6	262·6	279·6	272·5	290·2	296·2	314·9	321·6	340·5	359·1	5·0	22·7	27·7
7	274·8	291·8	284·7	302·4	308·5	327·2	333·8	352·7	11·3	17·2	34·8	39·8
8	287·0	304·0	296·9	314·7	320·7	339·4	346·1	4·9	23·5	29·4	47·0	51·9
9	299·2	316·2	309·1	326·9	332·9	351·6	358·3	17·1	35·7	41·6	59·1	64·1
10	311·4	328·4	321·3	339·1	345·2	3·8	10·5	29·3	47·9	53·7	71·3	76·2
11	323·5	340·5	333·5	351·3	357·4	16·0	22·7	41·5	60·0	65·9	83·4	88·3
12	335·7	352·7	345·7	3·5	9·6	28·3	34·9	53·7	72·2	78·1	95·5	100·5
13	347·9	4·9	357·9	15·7	21·8	40·5	47·1	65·9	84·4	90·2	107·7	112·6
14	0·1	17·0	10·0	27·9	34·0	52·7	59·3	78·1	96·6	102·4	119·8	124·7
15	12·2	29·2	22·2	40·0	46·2	64·9	71·5	90·3	108·7	114·5	132·0	136·9
16	24·4	41·3	34·4	52·2	58·3	77·0	83·7	102·5	120·9	126·7	144·1	149·0
17	36·5	53·5	46·5	64·4	70·5	89·2	95·9	114·7	133·1	138·8	156·3	161·1
18	48·6	65·6	58·7	76·5	82·7	101·4	108·1	126·9	145·3	151·0	168·4	173·3
19	60·8	77·8	70·8	88·7	94·9	113·6	120·3	139·1	157·4	163·2	180·6	185·5
20	72·9	89·9	83·0	100·9	107·1	125·8	132·5	151·2	169·6	175·3	192·7	197·6
21	85·0	102·0	95·1	113·1	119·3	138·0	144·7	163·4	181·8	187·5	204·9	209·8
22	97·1	114·2	107·3	125·2	131·5	150·2	156·9	175·7	194·0	199·7	217·1	222·0
23	109·3	126·3	119·5	137·4	143·7	162·4	169·1	187·9	206·2	211·9	229·3	234·2
24	121·4	138·5	131·6	149·6	155·9	174·7	181·3	200·1	218·5	224·1	241·5	246·3
25	133·5	150·6	143·8	161·8	168·1	186·9	193·6	212·3	230·7	236·3	253·7	258·5
26	145·7	162·8	156·0	174·0	180·3	199·1	205·8	224·5	242·9	248·5	265·9	270·7
27	157·8	174·9	168·1	186·2	192·5	211·3	218·0	236·8	255·1	260·7	278·1	282·9
28	170·0	187·1	180·3	198·4	204·7	223·6	230·3	249·0	267·3	272·9	290·3	295·1
29	182·1	199·3	192·5	210·6	217·0	235·8	242·5	261·3	279·6	285·2	302·5	307·3
30	194·3		204·7	222·8	229·2	248·1	254·8	273·5	291·8	297·4	314·7	319·5
31	206·5		216·9		241·4		267·0	285·7		309·6		331·6

Figure 9.6. A sample table from the *BAA Handbook* showing computed selenographic co-longitudes for the year 2000. *Courtesy The British Astronomical Association.*

noting the magnitude of the faintest stars that can be seen overhead. Ordinarily, a perfectly clear, dark sky will show stars as faint as magnitude +6 to the naked eye – though I've seen stars fainter than +7 from the high, dry unpolluted climates of West Texas and Northern Arizona. But light pollution, smog, and thin cloud will all reduce transparency.

Because the Moon is such a bright object, recording sky transparency as accurately as one might record it while observing faint objects such as galaxies is simply not necessary. But you should make note of any cloud or fog that substantially obscures or interferes with your ability to see the lunar features under observation. Passing clouds do not usually present too much of a problem to the lunar observer, as long as there are breaks in the cloud that allow you to clearly see the desired surface feature.

Seeing, which novice amateur astronomers often confuse with transparency, refers not to the sky's clarity but to the steadiness of the air between the observer's telescope and the object under observation. The more turbulent the air currents passing in front of the telescope, the worse the seeing – and the less that can be seen. Seeing can be affected by factors as diverse as upper air currents from a passing weather front, and heat waves rising from a driveway or rooftop. Never observe the Moon while it is over your or your neighbor's rooftop, especially after a particularly warm day.

If possible, place your telescope on grass, which will be much more thermally stable than the concrete or asphalt of a driveway or patio. If you are fortunate enough to be near a body of water, say a pond or lake, by all means set up your telescope next to it, since water is a great stabilizer of thermal currents. The best seeing conditions I ever experienced were in the Florida Keys and on Boca Grande, a barrier island off Florida's west coast. These observing sites were surrounded by water, and allowed me to attain the theoretical Dawes limit of resolution with my telescopes. Most observing sites aren't so ideal – unsteady air is usually the order of the day.

But how do we objectively quantify seeing conditions for the purposes of recording them on a lunar drawing? Historically, two different scales of seeing have been adopted by amateur observers of the Moon and planets: the Antoniadi scale and the Pickering scale. These two scales appear in Tables 9.1 and 9.2. The Antoniadi scale was named for Eugène Antoniadi (1870–1944), a longtime BAA member who became famous as a planetary observer during the first half of the twentieth century. The Antoniadi scale was specifically devised for lunar and planetary observers. On the Pickering scale, named for William H. Pickering, seeing of 1 to 3 is considered "very poor," 4 to 5 is "poor," 6 to 7 is "good," and 8 to 10 is "excellent." The Pickering scale requires the observer to looking at the diffraction rings round a star, which are most easily seen in refractors.

For annotating your lunar drawings, it really doesn't matter which scale of seeing you use so long as you choose one scale and stick with it, for consistency's sake. You should note that the two scales run in inverse order – a high number on the Pickering scale denotes excellent seeing, whereas a high number on the Antoniadi scale denotes very poor seeing. Recording the seeing on a regular basis allows you to calibrate lunar drawings made under different observing conditions, and is therefore one of the most important data entries you can make on your drawings. It does take practice to assess the seeing conditions accurately, but it will become easy to do after a while.

Table 9.1. The Antoniadi scale of seeing
1. Perfect seeing, without a quiver.
2. Slight undulations, with moments of calm lasting several seconds.
3. Moderate seeing, with large tremors.
4. Poor seeing, with constant troublesome undulations.
5. Very bad seeing, scarcely allowing the making of a rough sketch.

Table 9.2. The Pickering scale of seeing.
1. Image usually about twice the diameter of the third diffraction ring.
2. Image occasionally twice the diameter of the third ring.
3. Image of about the same diameter as the third ring, and brighter at the center.
4. Disk often visible; arcs (of diffraction rings) sometimes seen on brighter stars.
5. Disk always visible; arcs frequently seen on brighter stars.
6. Disk always visible; short arcs constantly seen.
7. Disk sometimes sharply defined. (a) Rings seen as long arcs. (b) Rings complete.
8. Disk always sharply defined. (a) Rings seen as long arcs. (b) Rings complete, all in motion.
9. (a) Inner ring stationary. (b) Outer rings momentarily stationary.
10. Rings all stationary. (a) Detail between the rings, sometimes moving. (b) No detail between the rings.

What to Do with Your Drawings

Once you have made a lunar drawing and recorded all the pertinent data, what should you do with it? There are at least two organizations that would be delighted to get a copy of your drawing, for their scientific archives, and for publication in their journals or newsletters. The British Astronomical Association, founded in 1890, has a Lunar Section that has been active since 1891. Lunar drawings sent to the BAA's Lunar Section may be published in either the *BAA Journal* or one of two publications of the Lunar Section itself. The Lunar Section *Circular* is a monthly newsletter that often contains drawings published by the Topographic Subsection. *The New Moon* is a quarterly magazine also published by the Lunar Section, and reproduces many drawings of lunar features.

Besides the BAA Lunar Section, you could also submit copies of your drawings to the Association of Lunar and Planetary Observers (ALPO), an American-based organization formed in 1947. ALPO publishes a well-illustrated journal called *The Strolling Astronomer*. Mass-market amateur astronomy magazines, such as *Sky & Telescope* and *Astronomy* in the USA, and *Astronomy Now* in the UK, occasionally publish lunar sketches. If you do submit copies of your drawings for publication, they should be high-quality photographic reproductions. Plain photocopies are almost never of sufficient quality for publication, though high-quality color photocopies (even if your original drawings are black and white) are sometimes good enough for the publisher to reproduce them.

Make sure you keep all of your lunar drawings on file, arranged alphabetically by the proper name of the feature drawn. You will find it helpful in improving your drawing skills to periodically return to a lunar feature you have drawn at some time in the past and compare the view through the eyepiece to your drawing. Make sure you are looking at the feature while it is at a similar co-longitude to that of your drawing. How realistic was your drawing? What improvements can you make the next time to portray the feature more accurately? Which aspects of your sketch are "dead-on" and which need

further work? You will also find that saving your drawings will allow you to detect subtle changes in the appearance of the same feature viewed at different co-longitudes. If the Sun angles are very different, the drawings may look nothing alike!

Now that you've made a survey of the Moon and completed your first drawings of lunar surface features, you may be ready to move on to more advanced lunar observing projects, such as photographing and imaging the moon with video cameras or CCDs, and observing eclipses and occultations. These topics are covered in the remaining chapters.

Chapter 10
Lunar Topographic Studies

In the previous two chapters we learned how to use the telescope to observe and draw lunar features. Once you have learned your way around the Moon's topography and acquired some proficiency in using the basic tools available to today's amateur lunar observer, you might start thinking about more advanced projects you can undertake.

I was recently chatting with a friend of mine who is a renowned lunar expert. He was one of the scientists chiefly responsible for the lunar mapping projects of the late 1950s through the mid-1960s that ultimately cleared the way for the Apollo astronauts to explore the Moon in safety from 1969 to 1972. When the subject of amateur lunar observing came up, he voiced an opinion I have heard many times over the years – that there is little left for the amateur to do here. I hope that this chapter will make a strong argument to the contrary: there *is* still much work for the amateur lunar observer to do.

In fact, my biggest challenge in writing the rest of this book was deciding what to select from the wide spectrum of amateur lunar research. There is now an absolutely astounding variety of interesting and worthwhile projects that amateurs all over the world are carrying out, some using time-honored traditional methods of observation, others taking advantage of new technologies that amateur astronomers could not have imagined when Apollo 11 landed on the Moon in July, 1969. In the end, I narrowed down the list – regrettably, but inevitably, leaving out many worthy projects. I have organized the projects described in the rest of the book into three principal categories: advanced topographic studies, including lunar transient phenomena (LTPs); lunar occultations and lunar eclipses; and imaging the Moon by photography, video, and CCDs.

There are many opportunities for the energetic lunar observer to seriously study the Moon's topo-graphic features. From the many advanced topographic studies that today's amateur lunar observers are involved with, I have selected three different programs: ALPO's Selected Areas Program, the Bright Lunar Rays Project (a joint project of ALPO and the BAA), and lunar transient phenomena (LTPs).

The Selected Areas Program

We have seen how to observe and draw lunar topographic features – but which ones to choose? There is no right or wrong answer to this question, of course. Some lunar observers concentrate on their own personal favorites, while others systematically survey the Moon, dividing it into various selenographic regions – by quadrants (as we did in Chapters 4 to 7), by phase, or by lunar latitude and longitude. Still other observers are fascinated by lunar geologic processes, and focus on specific formations that illustrate those processes. For example, if you are interested in lunar volcanic domes, you will sooner or later study the area near the crater Marius. If wrinkle ridges arouse your curiosity, you will want to survey the Mare Tranquillitatis and Mare Serenitatis regions. If you are having trouble deciding which features to observe, you may want to contribute observations and drawings to a very worthwhile effort of the Association of Lunar and Planetary Observers (ALPO) known as the Selected Areas Program, abbreviated to SAP.

ALPO was formed after World War II by famed Solar System observer Walter A. Haas and others who shared similar interests. In its first half-century of existence, ALPO has distinguished itself as a dynamic organization that has helped many amateur astronomers develop their skills as observers of the Sun, the Moon, the planets, comets, and meteors. These observers have repaid the investment by contributing to ALPO innumerable observations of all kinds, some leading to important discoveries or advances in our scientific knowledge and understanding of the Solar System. One of the major contributions to lunar science made by ALPO was the compilation of beautifully detailed maps for the region of the Moon known colorfully as Lunar Incognita, so called because, though visible from the Earth, this area was not mapped in good detail by the many spacecraft sent to the Moon during the late 1950s through the early 1970s.

In many ways, ALPO carries on in the USA a venerable tradition started by the British Astronomical Association (BAA), an organization that has long enjoyed great prestige for the high quality of observational work done by its members, especially on Solar System objects. Since 1890, the year of its founding, the BAA has published its own journal, known simply as the *Journal of the British Astronomical Association*. Happily, membership in both ALPO and the BAA is open to anyone who has a sincere interest in astronomy, and many amateur astronomers belong to both organizations. I strongly recommend joining both ALPO and BAA, whether you are a beginner or a more advanced observer.

Both organizations have many highly skilled and friendly expert observers to help beginners to acquire the knowledge and skills necessary to enjoy their new pastime. ALPO even has a special Lunar and Planetary Training Program, where novice observers receive a kit complete with special forms, instructions, and even drawing tools to help them get started. Participants in this program follow the step-by-step instructions provided in the kit and submit their work to expert observers who offer friendly advice on how to improve observing and artistic techniques, willingly sharing secrets it may well have taken them years of trial and error to learn.

Once you have acquired basic proficiency in drawing lunar features, either through a training program like ALPO's or on your own (perhaps by reading Chapter 9), you are ready for the next step – drawing selected lunar features. The advantages of participating in an organized observing program like SAP are many, including the opportunity to share and compare your results with those of many fellow observers. This collective effort creates a valuable database of observations that, taken together, are far more reliable than any one individual's data. Such a database can – and often does – serve as the basis for truly scientific papers. Inevitably, by participating in a program like SAP you will get to know other observers who share your interests – in this case lunar topography – which is one of the best thing about amateur astronomy.

What exactly is SAP? In previous chapters I have frequently mentioned that every feature on the surface of the Moon changes in appearance throughout the lunar month, as the phase angle of the sunlight falling upon it changes. Also, certain types of lunar feature tend to be brighter than others, for example the central peak of a large walled plain or crater, such as Aristarchus; other features tend to be darker, for example the floor of lava-flooded craters such as Plato. The brightness or darkness of these features is also affected by the passage of the lunar month and the Sun's changing phase angle, which we refer to as differences in "tone."

Certain lunar features, often the more complex ones, show variations in appearance that are unpredictable, perhaps simply because we haven't studied them closely enough to know how to predict their changes. A few of these more interesting lunar features have been selected by SAP for intensive, long-term study by amateur lunar observers like yourself. Besides keeping track of normal changes in the albedo, or reflectivity of light, for each of these features, SAP is also interested in documenting any unusual changes in appearance, including those that fall into the following categories:

1. *Tonal or color variations.* Look for radial bands or dark patches on crater floors, or a nimbus (halo) around the outside of the crater's walls. These aspects of the crater's appearance do not necessarily relate to changing phase angle in a predictable way.

2. *Shape and size changes.* We know that lunar features do not really change their size and shape, despite all the historical controversy and excitement over the crater Linné, for example. But they may *appear* to do these things, because of the changing phase angle of the Sun or libration effects, and it is important to document these apparent changes.

3. *Shadow anomalies.* Sometimes a shadow is not perfectly (normally) black, or may take on some unusual color or shape that can't be explained just by changing phase angle. (Some shadows will show lighter shadings within a jet black background. These shadings are occasionally caused by light scattered or reflected from other, nearby features, the light falling onto the floor of the crater where the shadow is projected.)

4. *Appearance or disappearance of features.* If you see any feature or detail of a feature that is not mentioned and in this or any of the classic lunar guidebooks, or does not appear on a high-quality lunar atlas, you should make a note of it and, if possible, obtain an image of it. Keep in mind that many of the features the classic observers claimed to see are simply not there (I have eliminated as many of these errors as possible from the descriptions in Chapters 4 to 7). If possible, check suspect features against a reliable, high-resolution image of the feature such as a Lunar Orbiter photograph or high-quality CCD image.

5. *Features exposed to earthshine.* Any anomalous tonal or albedo phenomena that are related to earthshine, the dim illumination of the night side of the Moon by sunlight reflected off the Earth. (Earthshine is a dimmer light than direct sunshine, and will therefore illuminate lunar features differently, with different shades of gray, compared to direct sunshine. Earthshine also falls upon lunar features from a slightly different incoming angle than direct sunshine, and this will also affect how the illuminated lunar feature appears to the observer.)

6. *Eclipse phenomena.* Any aspect of a feature in categories 1 to 4 above that is related to, or occurs specifically during, a lunar eclipses, and was not seen at any previous eclipse when the same feature was monitored.

What lunar features are included in ALPO's SAP? Currently the following craters are under study: Alphonsus, Aristarchus, Atlas, Copernicus, Plato, Theophilus, and Tycho. Data for these craters are presented in Table 10.1, showing the feature's name, selenographic latitude and longitude, and the Sun's co-longitude at sunrise, local noon and sunset. All of these craters are large and conspicuous, and have in the past displayed interesting changes in appearance. There is probably no better way to sharpen your lunar observing skills than to regularly scrutinize these complex and challenging features in successive lunations. To do this, SAP has adopted a standardized set of observing procedures:

1. Concentrate on one or two lunar features only throughout any given lunation. For each feature observed, you will be asked to assign values for the albedo (brightness) of several different index points located across the feature. How to do this is explained in detail below. Each observation should always be recorded on special forms provided by SAP, samples of which are on pages 170–176 and the CD-ROM .

2. You should carry out observations during a single lunation using the same telescope(s), eyepiece magnifications, and filters. If you are using more than one telescope, keep separate records for observations made with each telescope – do not mix them up! Keep in mind that many fine details visible in larger apertures will be invisible in smaller telescopes. If possible, apply the same consistency in your observing equipment and methods to successive lunations, for this makes it easier to compare data.

3. Maintain careful records of the date and time (use Universal Time, UT) for each observation, as well as the selenographic co-longitude (in degrees), and the orientation of the eyepiece's field of view (e.g., south up, west to the right). Record also the sky transparency and seeing conditions (see Chapter 9). The SAP forms have space for all of this information.

Table 10.1. Data for craters in the ALPO Selected Area Program

Crater	Selenographic		Sunrise co-long.	Local noon co-long.	Sunset co-long.
	Lat.	Long.			
Alphonsus	4°W	13°S	4°	94°	184°
Aristarchus*	47°W	23°N	47°	137°	227°
Atlas	43°E	46°N	317°	47°	137°
Copernicus	20°W	9°N	2°	110°	200°
Plato	9°W	51°N	9°	99°	189°
Theophilus	26°E	11°S	334°	64°	154°
Tycho	11°W	42°S	11°	101°	191°

*Herodotus, because of its proximity to Aristarchus, is included in this composite SAP feature.

4. Observe the selected features only when the Moon is at an altitude of 25° or greater above the horizon. Otherwise, poor seeing conditions and the dispersive effects of the atmosphere will interfere with your observations to an unacceptable degree.

5. SAP provides "Assigned Albedo Index Points" charts (see pp. 170–176 and the CD-ROM) for each selected feature. On each chart, several "index points" are plotted and assigned letters; the observer then estimates the albedo at each point:

   (a) the principal cardinal points (compass directions), according to the International Astronomical Union (IAU) convention, for the inner slopes of crater walls and the outer slopes of crater mountains or domes;

   (b) the summit of the crater's central peak(s), and any other significant crater mountain;

   (c) features of the floor of the crater or on the nearby surrounding terrain.

So that results from different observers (and, for that matter, different observations by the same observer) may be consistently compared, you must take care always to use the index points as they are assigned. If you decide to estimate albedo values for other, non-indexed features, keep the data separate and make careful notes and drawings to indicate the precise location of the non-indexed features you are observing. But how do you go about estimating albedo values?

First, you need to obtain a reliable "gray scale wedge," an example of which is shown in the first column of Table 10.2. This is just a standardized scale showing various "shades of gray" from white to black. You should be able to find such scales at an art supply, photography, or printing shop. Or you can make your own, by shading a white strip of heavy paper or cardboard with bands of gray using standard artist's pencils. Other observers have made such scales by exposing pieces of black and white 35 mm film to different densities. You will need at least ten different gradations or steps, preferably twenty. Next, take your scale to the telescope when the Moon is up, and refer to Table 10.2. This table gives what is called Elger's albedo scale, in honor of the English amateur lunar observer who invented it. Standard albedo values, on a scale of 1 to 10, are assigned to various features easily located on the lunar surface. Where possible, Table 10.2 gives more than one example of features having the reference albedo, so that you will be more likely to find at least one that is visible on any given night.

Next, to calibrate your gray scale you must observe at least one feature, if possible, for each of the standard numerical values listed on Elger's scale. They should match up pretty closely with the standard 1–10 scales that are commercially available, but keep in mind that you may see the features on Elger's scale a little differently. What is important is to establish your own personal scale of albedo, so that you can consistently assign values the way *your* eyes detect different shade of gray. You must use the same telescope when calibrating the gray scale that you will use to observe SAP features.

Once you have calibrated your scale against Elger's, pick one of the SAP features for observation and begin looking at each of the index points labeled on the SAP reference chart for that feature. Assign a value for albedo from your gray scale for each index point, simply by matching its perceived brightness (or darkness) to the gray scale. Record the data in the column marked "Albedo IL." This means albedo for "integrated light", which simply indicates that you are not using any filters. Try to illuminate the gray scale with the same light as the light of the Moon in the eyepiece. In going through this exercise you are simply looking at different parts of a crater and trying to decide which part of your gray scale it resembles most closely. This takes practice and patience, and at all times you must be careful that you are in fact looking at the right part of the crater, mountain peak or other feature as is indicated by the index points. This can be more difficult than you might think.

Throughout a lunation, you shouldn't notice huge changes in the north and south IAU points for any given feature, for they will, at any given time, be receiving approximately the same amount of sunlight. However, this is not always exactly true – one point might be in deeper shadow than the other if the Sunlight to its east or west is being blocked by a high peak, for example. Or one of the points might get "early" sunlight through a gap in the crater rim, which is usually a series of interrupted peaks.

You will likely notice much more variation throughout the course of a lunation in the albedos of IAU east and west points for a given feature, the east wall of a crater looking dull at sunrise while the west wall catches the first rays of sunlight, the opposite effect occurring at sunset. As the sun rises higher, the east wall will brighten while the west wall grows duller, until sunset, when the east wall will be at its brightest, the west at its most obscure. The crater floor will normally brighten until local noon, when the Sun is directly overhead the feature, then darken again as the Sun gradually sets. This will obviously have a big effect on the visibility of features on the floor.

Table 10.2. Elger's albedo scale and a gray scale wedge

Gray scale	Standard Albedo value	Corresponding lunar features
	0.0	Black shadows
	0.5	None
	1.0	Darkest parts of Grimaldi, and Riccioli
	1.5	Interiors of Billy, Boscovich, and Zupus
	2.0	Floors of Endymion, Le Monnier, Julius Caesar, Cruger, and Fournier
	2.5	Interiors of Azout, Vitruvius, Pitatus, Hippalus, and Marius
	3.0	Interiors of Taruntius, Plinius, Theophilus, Parrot, Flamsteed, and Mercator
	3.5	Interiors of Hansen, Archimedes, and Mersenius
	4.0	Floors of Manilius, , Ptolemaeus, and Guericke
	4.5	Surfaces around Aristillus and Sinus Medii
	5.0	Walls of Arago, Lansberg, Bullialdus; surfaces near Kepler and Aristarchus
	5.5	Walls of Picard and Timocharis; rays of Copernicus
	6.0	Walls of Macrobius, Kant, Bessel, Mösting, and Flamsteed
	6.5	Walls of Langrenus, Theaetetus, and La Hire
	7.0	Theon, Ariadaeus, Bode B, Kepler, and Wichmann
	7.5	Ukert, Hortensius, and Euclides
	8.0	Walls of Godin, Bode, and Copernicus
	8.5	Walls of Proclus, Bode A, and Hipparchus C
	9.0	Censorinus, Dionysius, Mösting A, and Mersenius B and C
	9.5	Interiors of Aristarchus and La Pérouse
	10.0	Central peak of Aristarchus

Besides the charts showing albedo index points, there are "Drawing Outline Charts" for each of the SAP features. These charts contain sufficient details already plotted to enable you to accurately sketch in additional, finer details not shown on the charts. This is a great exercise for training your eye to appreciate subtle details, and is important for recording any perceived changes in size or shape of the details that correspond to albedo index points, as well as any other details of interest that you notice.

Obviously, this part of the exercise is fairly subjective, so it requires more experience as an observer to yield meaningful results. This is also where you should indicate the tone of fine details. Use the form provided for recording your "Albedo Values and Supporting Data for Lunar Drawings", and the one drawing in fine details, headed "Visual Observations of Selected Lunar Features" (see the CD-ROM). These forms also prompt you to include the necessary details of your telescope and conditions under which you observed.

Notice that the second of these forms has extra columns for observations made with filters ("Albedo: Filter 1," "Albedo: Filter 2," etc.), which we discussed briefly in Chapter 8. These filters usually consist of Kodak Wratten gelatin film sandwiched between two pieces of glass, mounted in metal rings that screw onto the end of your eyepiece. These wonderful filters are usually very affordable, and have the valuable property of transmitting only specified wavelengths of light. Make sure that any filters you acquire indicate the wavelength of light they transmit – some inferior filters fail to do this. The SAP has adopted the following Wratten filters for use when observing its selected features: W23A or W25 (red light), W38A or W47 (blue light), and W58 (green light). Use the W23A and W38A filters for small telescopes, as these are less dense than the others and will transmit more light. Using these filters, re-examine your selected lunar feature and repeat the albedo exercise described above. Be sure you indicate the Wratten number of the filter for each albedo measurement you make.

The "Visual Observations of Selected Lunar Features" form on page 174 (and on the CD-ROM) also has blank spaces where observers can draw albedo reference and outline charts for lunar features of their own choosing. Use this form when you wish to submit observations for any feature not listed in Table 10.1. You are encouraged to contact the ALPO Lunar SAP Recorder (see "Resources for the Lunar Observer" on the CD-ROM) to purchase a copy of the *Lunar SAP Handbook*, which contains a list of additional lunar features for monitoring.

Each SAP observing form contains a space for the observer's notes. What should you write here? You should make careful notes of any features that are visible or obvious only under low or high Sun angles, as well as the nature and extent of bright rays, dark bands, and the general appearance of the feature and its surrounding environment. Although the primary aim of the SAP is to record changes that take place throughout a lunation or during librations, you should also note any short-term changes that appear to take place during the course of the observing session. Make copies of your observing forms and send the originals to the ALPO Lunar SAP Recorder.

Finally, SAP provides a "Dark Haloed Craters Observing Form" and a "Bright and Banded Craters Observing Form" (these are also on the CD-ROM). Dark haloed craters, or DHCs, were once thought by many to be volcanic in origin. You might recall from the discussion of lunar geology in Chapter 3 that in the days before it became apparent that the impact of asteroids was the dominant crater-forming process on the Moon, many lunar scientists thought that the lunar craters were volcanic in origin. Now we know that lunar craters are not volcanoes, but impact sites. Large-scale lunar volcanism likely ceased 3 billion years ago. Unfortunately, many amateur astronomers did not keep up with advances in lunar geology, and continue to believe that many lunar craters are volcanic.

The DHCs are young, usually very small, craters that are surrounded by a halo of mysterious dark material instead of bright rays. DHCs may be found on the floor of Alphonsus and near Copernicus. It has long been suggested they are cinder cones or fumaroles, which is a reasonable interpretation given their appearance and their frequent association with lunar rilles, some of which are probably lava tubes. Apollo 17 explored a DHC, dubbed

"Shorty," which was found to be a regular impact crater with no evidence of volcanism. The dark halo was not volcanic cinders, but orange and black glass, formed when rock was melted by the intense heat resulting from impacts, and then resolidified, casting further doubt on the volcanic origin of DHCs. Nonetheless, these unusual formations continue to fascinate amateur observers, and the *Lunar SAP Handbook* has a list of them for monitoring. It also contains a very interesting discussion of the results of the DHC sub-program to date.

The "Bright and Banded Craters" form is for use when monitoring craters that are extremely brilliant when the Sun is overhead (near full moon), and craters that show dark or light radial bands within their walls. Lunar geologists attach no particular significance to these features, but certain craters do seem to show them, and they may be telling us something about slumping and other post-impact modification processes that have gone on inside lunar craters since they were created. At one time it was very fashionable for the classic lunar observers to observe these bands and catalog them, sometimes using these phenomena to support wild theories that we now know are geologically unsound.

But the light plays many tricks on the Moon's surface, and it is both fun to observe such albedo features and to use them to sharpen your observing eye. Certainly some lunar craters are much brighter than others at high Sun angles, and some of the brightest craters with the most impressive ray systems are not necessarily the biggest or deepest, a condition which offers no ready explanation. The *Lunar SAP Handbook* contains a list of bright and banded craters, describing in detail the objectives of this sub-program. If you join ALPO you will have the opportunity to participate in other, similar programs, including the discovery, cataloging, and observation of the puzzling features known as lunar domes, which may be the Moon's equivalent of terrestrial shield volcanoes.

In 1999, ALPO and the BAA announced that they were collaborating on a new, exciting project that any amateur lunar observer can participate in. Known as the Bright Lunar Rays Project, this unique program seeks to study an aspect of the lunar surface that has long been overlooked. Yet there can be no doubt that the lunar rays are a very important clue to the precise details of the impacts that created lunar maria basins and craters, nor can it be said that they are completely understood, even by professional lunar scientists. So they are certainly worthy of further study, and who better to do this than amateur lunar observers? The following description of the Bright Lunar Rays Project was written by ALPO's Coordinator for Lunar Topographical Studies.

🌙 ○ 🌙 ●

The Bright Lunar Rays Project

Bill Dembowski

Conventional wisdom says that the worst time for lunar observing is when the Sun is high in the lunar sky. It is at these times that shadows disappear and lunar features are difficult, if not impossible, to observe. But conventional wisdom does not always apply. When the Sun is high in the lunar sky some of the Moon's most fascinating features, the bright lunar rays, blaze into view.

Lunar rays, those beautiful splash patterns that cover the face of the full moon, were once quite a mystery. Some early observers thought that they were cracks in the lunar crust that were later filled by dust or ice; others considered them to be salt deposits from extinct oceans. One of the more persistent theories was that they were volcanic in origin, similar to the features known as Pele's Hair in Hawaii. We know now that they are the ejecta of meteoric and asteroidal impacts.

Rays appear to be distributed randomly across the lunar disk, though those on the dark maria tend to show up most dramatically because of the contrast effect. Although often quite extensive, they have no appreciable height and are never seen to cast a shadow. They are mostly uninterrupted by mountains, crater walls, or rilles, but there is some observational evidence that this is not always the case.

Table 10.3 lists the lunar coordinates for 35 craters having prominent ray systems, with Reiner γ thrown in for good measure. In addition to conventional ray systems there are literally hundreds of bright spots on the lunar surface. These are, of course, simply rays of limited extent. A listing of these bright spots would have to be made primarily on the basis of their coordinates since most of the craters are so unremarkable as not to have names. Rays themselves, unlike all other major features on the Moon, do not have a standard nomenclature to identify them.

Rays are best seen under high illumination. It is advisable to begin your observing session with a rel-

Table 10.3. Coordinates for selected bright ray craters

Crater	Selenographic	
	Lat.	Long.
Anaxagoras	73°.4N	10°.1W
Aristarchus	23°.7N	47°.4W
Aristillus	33°.9N	01°.2E
Autolycus	30°.7N	01°.5E
Bessel	21°.8N	17°.9E
Birt	22°.4S	08°.5W
Byrgius A	24°.5S	63°.8W
Copernicus	09°.7N	20°.0W
Euclides	07°.4S	29°.5W
Furnerius A	33°.5S	59°.1E
Geminus C	33°.9N	58°.7E
Glushko	08°.3N	77°.5W
Godin	01°.8N	10°.2E
Hind	07°.9S	07°.4E
Kepler	08°.1N	38°.0W
Lalande	04°.4S	08°.6W
Langrenus	08°.9S	60°.9E
Manilius	14°.5N	09°.1E
Menelaus	16°.3N	16°.0E
Messala B	37°.4N	59°.8E
Messier A	02°.0S	46°.9E
Olbers	07°.4N	75°.9W
Petavius B	19°.8S	57°.1E
Proclus	16°.1N	46°.8E
Reiner γ	08°.0N	58°.0W
Sirsalis	12°.5S	60°.4W
Snellius	29°.3S	55°.7E
Stevinus A	32°.1S	51°.9E
Strabo	61°.9N	54°.3E
Taruntius	05°.6N	46°.5E
Thales	61°.8N	50°.3E
Theophilus	11°.4S	26°.4E
Timocharis	26°.7N	13°.1W
Tycho	43°.3S	11°.2W
Zucchius	61°.4S	50°.3W

ative low-power eyepiece (×20 to ×50) to get an overall view of the larger ray systems and their relationship to one another. Move next to a medium power (×100 – to ×200) to study individual systems. Some observers, however, find that magnifications in excess of ×200 are more of a hindrance than a help when studying the rays. Also, when you are observing the brighter regions of the Moon you will usually need to take some measure to reduce glare. Reducing the aperture of your telescope is one way of doing this, but most observers prefer the use of filters – either colored, neutral density, or polarizing. With such a filter in place, a few rays can actually be traced nearly to the terminator.

The most extensive lunar ray system is that associated with the crater Tycho. Long and straight, the rays of Tycho reach halfway across the face of the Moon and are so bright that they are highly visible even in the relatively bright highlands where they originate. One ray from Tycho appears to divide Mare Serenitatis in half. If this is in fact a ray from Tycho – and not from Menelaus, as some believe – then the span of the Tycho ray system would exceed 2000 km (1250 miles). Interestingly, the Tycho rays do not begin at the crater walls. There is a dark halo surrounding the crater which may be the result of darker, heavier ejecta piling up around the crater walls. It should also be noted that the Tycho ray system is far from symmetrical. Some of its longest rays do not radiate from the center of the crater but from points on the crater walls. In addition, there is about a 120° gap in the system to the east. This gap is not totally devoid of rays, but it lacks the major streaks that typify the rest of the system.

To the southeast of Tycho are the craters Copernicus and Kepler. Their ray systems, though prominent, are rather different than Tycho's. Whereas the rays of Tycho are straight and narrow, these are broader and more feather-like. They seem to spread like ostrich plumes, sometimes doubling back upon themselves to form oval loops. The two systems often overlap in a complex pattern that is difficult, if not impossible, to decipher. Both of these ray systems are also dimmer than that of Tycho, an indication that they predate their southern neighbor. It is generally accepted that rays are bright when originally formed but darken as they are bombarded by micrometeorites and cosmic rays. One puzzling aspect of the Copernicus rays is that they are brighter than their age suggests they should be. At 810 million years old, theory says that they should have faded by now but at full moon they are second only to Tycho in brilliance and extent.

To the south of Copernicus and Kepler is Aristarchus. Recognized as the brightest crater on the Moon, Aristarchus is not often thought of as the center of a ray system. Inspection of the area surrounding Aristarchus reveals a dim but definite ray system, complex in nature and quite widespread. Obviously there is a relationship between the brightness of a crater and the existence of rays around it.

One of the most interesting ray systems is the one associated with the crater Proclus on the eastern shore of Mare Crisium. Proclus, incidentally, is the Moon's second brightest crater. This system has a most definite gap of 180° that completely spares Palus Somni from any intrusion. There is some observational evidence to suggest that a ridge in the area interrupted the trajectory of the low-flying ejecta, but the acceptance of this explanation is not universal. Regardless of the cause, this gap in the ray pattern is an echo, in miniature, of the Tycho system.

The pair of rays emanating from the twin craters Messier and Messier A are truly unique, as they stretch comet-like from the center to the western shore of Mare Fecunditatis. Closer inspection will reveal that both rays actually emanate from only one of the twins, Messier A (formerly known as Pickering). Current thinking is that both of the craters, and the rays, were formed by the glancing impact of a meteorite that struck the lunar surface at the incredibly shallow angle of 5°. These rays are visible even under a relatively low Sun angle and make for interesting viewing anytime they move off the terminator.

On the extreme western shore of Oceanus Procellarum is one of the most extraordinary features on the Moon, Reiner γ. Reiner γ is a swirl of bright ray-like material that has no known source. It appears brighter under a low Sun and, like lunar rays, it never casts a shadow. But where did it come from? A similar feature exists on Mare Marginis on the Moon's eastern limb. It has been theorized that it is the result of ejecta emanating from the impact that formed Mare Orientale and converging from all directions on the exact opposite side of the Moon. But no such impact site exists to account for the presence of Reiner γ. Is it, in effect, a ray without a crater to call home, or a feature in a class by itself?

Figure 10.1 is a photograph taken for the Bright Lunar Rays Project. The purpose was to show the many rays and bright spots on the surface of Mare

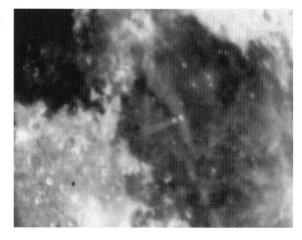

Figure 10.1. A photograph of Mare Fecunditatis taken for the Bright Lunar Rays Project. *Photograph by Bill Dembowski.*

Fecunditatis, particularly those in the vicinity of the Messier twins. The telescope used was a 125 mm (5-inch) *f*/9 apochromatic refractor equipped with a ×2 Barlow lens, an Astrovid 2000 camera (see Chapter 12 for more on the Astrovid), and a high-resolution monitor. The camera is capable of resolving 600 lines and the monitor, at 700, is able to show all that the camera can see. A standard VHS VCR, however, can resolve only 200–250 lines – a significant loss of quality.

As an inexpensive alternative I have been experimenting with photographing the monitor screen with fine-grain film, and the results have been quite satisfactory. With this setup I can have the advantages of both video and film. Focusing is simpler, and shutter speeds as short as $\frac{1}{16}$ s. are possible. Since there is no physical contact between the film camera and the telescope, there is no shutter vibration. Also, on a more personal note, as a lifelong darkroom enthusiast I prefer a photographic print to an image printed off a computer.

Bright lunar rays are so intriguing that the Association of Lunar and Planetary Observers, the British Astronomical Association, and the American Lunar Society are conducting a joint program to catalog, map, and study them in detail. Anyone wishing to participate in the program should contact the ALPO Coordinator of Lunar Topographical Studies (see "Resources for the Lunar Observer" on the CD-ROM).

Lunar Transient Phenomena (LTPs)

No book on amateur lunar studies would be complete without at least a brief discussion of lunar transient phenomena, a controversial subject to say the least. For centuries, lunar observers have reported strange happenings on the Moon – lunar features appearing to change size or shape, or to show anomalous color, bands of light or dark shimmering across the insides of craters, or eerie but fleeting glows of light or obscuring clouds suddenly appearing out of nowhere. I have always been surprised by the great number of these reports, for in nearly thirty years of observing the Moon I have never seen anything out of the ordinary, though I have enjoyed watching the Sun play many tricks with the light as it rises and sets over the lunar landscape.

A strong case is made against the reality of most LTPs by two amateur astronomers and authors, William Sheehan and Thomas Dobbins, in an article published in the September, 1999 issue of *Sky & Telescope* magazine. Titled "The TLP Myth: A Brief for the Prosecution," the article calls into serious question the reliability of two of the more famous "documented" cases of LTPs – Nikolai Kozyrev's 1958 reports of volcanic activity in the crater Alphonsus, and the so-called Aristarchus glows.

Kozyrev obtained a spectrum of an LTP that supposedly took place over the central peak of Alphonsus – according to Kozyrev, the spectrum showed emission bands toward the blue end of the spectrum. Many professional astronomers who examined the spectra concluded that the bands were really artifacts of faulty "guiding" – inaccuracies in the telescope's tracking of the fast-moving Moon. The results were never duplicated, and duplication of results is the cornerstone of the scientific method. It was also pointed out by noted planetary scientist Gerard P. Kuiper that Kozyrev's "bands" should have showed up as dark absorption bands against the brilliant white peak of Alphonsus, not as the bright "emission" bands that Kozyrev thought he saw. Also, consider this: despite the literally millions of images of the Moon taken by spacecraft since the early 1960s, no LTP has ever been "caught in the act."

Many amateur LTP enthusiasts wonder why more professional lunar scientists are not more interested in LTPs. Although there are always at least one or two "poster papers" presented at planetary science conferences on LTPs, some of them quite interesting, lunar scientists tend not to worry themselves about this question. The main reason for this is that LTPs, assuming that they are observable from Earth (recall that even the July, 1999 Lunar Prospector impact was not observable from Earth, though we had powerful telescopes with highly sensitive detectors trained on the impact site) are almost certainly nothing more than minor "out-gassings" of radon or other gases that do little if anything to affect the lunar environment, much less reshape the lunar surface.

We know today from the Apollo and Clementine lunar geologic studies that large-scale lunar volcanism ceased 3 billion years ago. Lunar geologists are more interested in questions surrounding the Moon's origin and its evolution into its present state than they are about current small-scale events that have no bearing on those larger questions. Professional scientists are forced to budget their grant monies effectively, and devoting resources to studying LTPs is not likely to pay big dividends, as far as our knowledge of the Moon's origin or geologic history is concerned. That is not to say that interesting things do not happen on the Moon today, for we know that small meteorites continue to pelt the Moon, weathering its features over geologic timescales, and it is certainly possible that LTPs do occur. But these events pale in comparison with giant impacts, multi-ring-forming impacts, magma oceans, and mare-filling volcanism.

Having said all that, I would tend to agree with a professional lunar scientist friend of mine who asked, "If interest in LTPs get amateur astronomers to look at the Moon, what's wrong with that?" For the amateur lunar observer who is interested in LTPs, both ALPO and the BAA will gladly accept reports of any anomalous activity that you observe. Figure 10.2 shows a completed LTP report form of the type accepted by the BAA's Lunar Section.

```
ROUTINE REPORT/TLF REPORT

DATE.....1980 - 7 - 23......

NAME... A Observer............ ADDRESS. The Observatory,. Newtown,. Kent......
                                .....................................
                        TEL. No. 123 567.............................

INSTRUMENT USED (type & size)..12". Newtonian Reflector. x 180. x240.........

SEEING CONDITIONS. Variable III to II . Transparency fair, some turbulence....
(Antoniadi scale plus indication of transparency)

LOCAL CONDITIONS Calm, slight mist, Temp 63°F. Barom. 30.1.................

ACCESSORIES USED CED, blink, camera.......................................
(i.e. Blink-CED- Photo-Photometer etc.)

SPURIOUS COLOUR. General spurious colour, much in evidence in proximity terminater
(please indicate extent and regions affected- alternately state if no colour seen)
********************************************************************************

Active from 20.30 until 21.50 UT              All east - west direction IAU

Aristarchus  20.45 UT Terminater just clearing area, west wall partially illuminated
             but interior of crater still shadowed. On the western wall, south west
             inner corner two craterlets seen also radial bands on this wall still
             contain shadow. Both eastern outer and western inner walls heavily
             tinged with spurious colour, blue.  The craterlets on the west wall
             tend to merge into background when seeing falls to III.
             Beyond the terminater in the
             darkened zone, approximately
             50 miles to the north west of
             Aristarchus a high peak is          small sketch of
             illuminated by the sun.             Aristarchus


Photograph taken at 20.48 UT

20.50 to 21.10 UT Blink  survey
areas scanned, Alphonsus, Theophilus, Aristarchus, Bullialdus, Plato and
Fracasterious. No response other than the permanet blink in Fracasterious
south west quadrant, usual brightening in red filter.

Gassendi  21.15 UT Much spurious colour over region, in particular intense
          redness along top of eastern rim.
          Some shadow remains along inner
          eastern wall, extends well onto        small sketch of
          floor at position between seven        Gassendi
          and ten o'clock, Shadows still
          produce much relief central peaks
          and crater floor.

21.23 to 21.45 UT CEE values.
                        Piton...... 2.7
                        Pico....... 3.1      Five readings each all
                        Tycho...... 3.4      constant, no varaition
                        Bullialdus. 3.6
                        Gassendi... 3.2
                        Aristarchus
                          west wall.1.9

REGIONS SCANNED/REGIONS OBSERVED/TLP CONDITION ALERT  (please indicate type of
                                                             observation)
```

Figure 10.2.
A completed LTP report of the type accepted by the BAA's Lunar Section. *Courtesy The British Astronomical Association.*

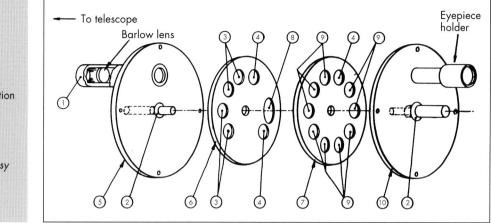

Figure 10.3.
A "crater extinction device" for accurately measuring the albedo of lunar features. *Courtesy The British Astronomical Association.*

Obviously, getting a video or CCD image of an LTP is the preferred method of documenting what you see.

As for LTP observing techniques, they are basically the same as those employed for observing any lunar surface feature, described in Chapters 8 and 9. Many participants in LTP programs will routinely examine specific lunar features for evidence of activity, such as color abnormalities, variations in albedo (brightness or reflectivity), obscurations, or flashes of luminescence. Some of the lunar features better known for generating reports of LTPs are Alphonsus, Aristarchus, Bullialdus, Cape Laplace, Daniell, Gassendi, Grimaldi, Manilius, Menelaus, Mt Pico, Plato, the Spitzbergen Mountains, Torricelli B, Theophilus, and Tycho.

The BAA's *Guide to Observing the Moon* includes a very detailed chapter on how to look for LTPs, and contains instructions for building a "crater extinction device" for accurately measuring the albedo of lunar features. Figure 10.3 shows this device, which allows the observer to combine neutral density filters (available at any camera shop) so that at some point the lunar feature under observation will be "extinguished." The total density of combined filters required to extinguish the feature will correspond to its albedo – bright features will require a greater total density of filters before they disappear than will darker features. It should be obvious that the crater extinction device is very useful for any lunar studies where accurately measuring the albedo of surface features is important, such as the SAP described earlier in this chapter, so it is a valuable tool that I highly recommend to any lunar observer.

Chapter 11

Lunar Eclipses and Occultations

Because the Moon is constantly in motion, it will, from time to time, grab the keen observer's attention by involving itself in interesting celestial phenomena, total solar eclipses being the most obvious example. Because there are many good books on solar eclipses(including one in this series – Michael Maunder and Patrick Moore's *The Sun in Eclipse*), and because total solar eclipses are events for solar rather than lunar observers, I shall discuss them no further here. But what of lunar eclipses? These phenomena, which are very pleasing to look at, also offer the amateur lunar observer the chance to do interesting scientific experiments. For some reason this aspect of lunar eclipses is often ignored. We shall try to do something about that in this chapter.

Another category of lunar studies that should interest the serious observer is the occultation, or eclipse, of bright stars by the Moon, especially the so-called *grazing occultations*, when the Moon's northern or southern limb barely eclipses the star. Both the BAA and ALPO support these studies, which do not require sophisticated or particularly expensive equipment. We shall take up occultations in the second half of this chapter.

Lunar Eclipses

Undoubtedly, human beings have observed lunar eclipses with the naked eye for as long as the human race has existed. Johannes Kepler was probably the first astronomer to observe a lunar eclipse through a telescope, in the year 1610. These events, things of beauty to us today, were regarded with terror by ancient and even medieval peoples. The typical deep red coloration of the totally eclipsed Moon was too frequently equated with bloodshed by fearful persons who did not understand how natural and harmless was this celestial happening.

Figure 11.1 illustrates the aspects of a lunar eclipse. As we learned in Chapter 1, a lunar eclipse will occur whenever the Moon, in its orbit around the center of gravity of the Earth–Moon system, moves into the shadow of the Earth, either wholly or partly. Figure 11.1 shows that Earth actually has two zones of shadow, a narrower, darker zone known as the umbra, and a wider, lighter one we call the penumbra. The penumbra is actually slightly larger, by about 2%, than predicted by the pure geometry of the Sun–Earth–Moon system, due to our planet's thick atmosphere, which diffuses sunlight by scattering and refraction.

The umbra is about 1° 22′ wide, the penumbra 2° 26′ wide. At the average Earth–Moon distance the umbra is 9180 km (5,705 miles) across, more than twice the Moon's diameter. A total umbral eclipse of the Moon occurs when the Moon moves completely into the umbral zone. Because red light is preferentially refracted by Earth's atmosphere at sunrise and sunset, the umbral shadow is not pitch black – instead, this refracted sunlight illuminates the totally eclipsed Moon with a light that is anywhere from 0.02% to 0.00 002% of the normal brightness of a full moon. If you were observing a total lunar eclipse from the Moon instead of the Earth, you would see the Sun covered by a dark Earth, our planet surrounded by a ring of red refracted light. Partial umbral eclipses, where the Moon moves partly, but not totally, into the umbral zone, are also possible. Aristotle was able to deduce that the Earth was a sphere from his observation of partial umbral eclipses – the great philosopher noticed that the Earth's shadow cast on the Moon during these events was always in the form of an arc. Figure 11.2

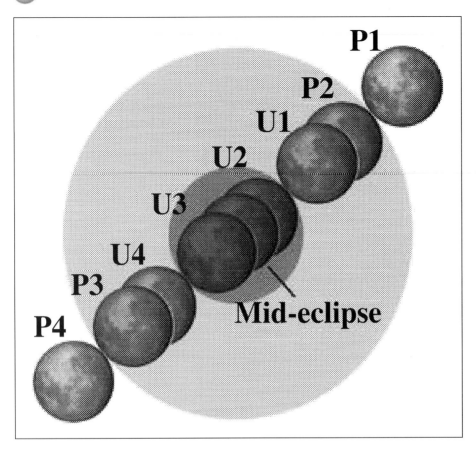

Figure 11.2. The Moon entering the umbral eclipse phase. *Photograph by Bernard Wahl.*

is a photograph of the Moon entering the umbral phase of the eclipse.

The penumbral shadow is lighter as a consequence of simple geometry – some sunlight is able to get by the Earth if the Earth is not perfectly interposed between the Sun and the Moon in a straight line. But the alignment has to be close to perfect or there is no eclipse at all, so lunar eclipses always occur at or very near full moon. A penumbral eclipse occurs when part or all of the Moon enters the penumbra, but no part of it enters the umbra. Unless you are aware that a penumbral eclipse is in progress, you might not even notice it, as the only effect is a slight dimming of the full moon that is not always obvious, and a deeper gray duskiness in the direction of deepest penetration by the Moon into the penumbral zone. In fact, you will not be able to perceive the penumbral shadow outside of the inner 60% of the penumbral zone.

For the amateur lunar observer, total umbral eclipses are both the most spectacular and the most scientifically interesting type of lunar eclipse to study. A total umbral eclipse, which for simplicity we shall refer to as a "total lunar eclipses," proceeds through several stages, as indicated in Figure 11.1:

P1   *First penumbral contact*, where the Moon first makes contact with the penumbral zone. The start of this stage is generally unobservable. About 30 min into this stage of the eclipse you should start to notice a slight duskiness on the Moon's western limb.

P2   *Second penumbral contact*, where the Moon is just wholly inside the penumbral zone. The whole Moon is now dusky.

U1   *First umbral contact*, where the Moon first contacts the umbral zone. Careful observers will be able to detect this event immediately after the Moon reaches U1, as a noticeable darkening will occur after that point.

U2   *Second umbral contact*, where the Moon is entirely within the umbral zone; *totality* begins at this point. As the eclipse progresses from U1 to U2, look for the two different shadows – the umbral and penumbral, separated by a sharp arc. The outer umbra may appear bluish or pearly-white. Just before U2 is reached, the penumbral shadow narrows to a thin ring, then vanishes as "totality" begins.

Mid-eclipse   *Maximum eclipse* (minimum lunar brightness) occurs at this point, half-way between U2 and U3. The un-eclipsed full moon shines at –12.7 visual magnitude; during mid-eclipse this value ranges from –3.0 to +4.0. The totally eclipsed Moon is usually a deep red at this point.

U3   *Third umbral contact*, the last moment at which the Moon is entirely within the umbral zone, basically a mirror image of U2. After this point the penumbral shadow gradually re-appears, first as a very thin ring around the umbral shadow.

U4   *Fourth umbral contact*, where the Moon just loses contact with the umbral zone and is now fully within the penumbral zone. The Moon brightens considerably as it leaves the umbral zone.

P3   *Third penumbral contact*, just before the Moon begins its departure from the penumbral zone. The Moon has lost all of its darker shadowing, including its red color, but the whole disk is covered by a dusky light, as in P2. After this point part of the Moon is in the penumbral zone, and part of it is outside the penumbral shadow.

P4   *Fourth penumbral contact*, when the Moon finally departs the penumbral zone entirely; the eclipse is over after this point. Before it reaches P4, only the east limb of the Moon shows a slight duskiness.

The magnitude of a lunar eclipse is defined as the degree to which the Moon is within the umbral zone, expressed by the ratio $d/0°.5$, where d is the angular distance inside the umbral zone (the distance from the outer edge of the umbral zone to the leading edge, or western limb, of the Moon's disk) and $0°.5$ is the Moon's apparent angular diameter as seen from Earth. For total lunar eclipses this ratio ranges from 1.0 to 1.8 (at mid-eclipse); for partial umbral eclipses it varies between 0 and 1.0. Totality begins when the magnitude of the eclipse is 1.0.

But what kind of useful work can the amateur lunar observer do during lunar eclipses? One activity that I enjoy is *crater timings* – very precise measurements of the times when particular craters enter and exit the umbral shadow during a total lunar eclipse. To perform crater timings, you first need to wait until the umbral phase begins, then center your telescope on the umbral boundary. Pick out an easily observed, preferably large, lunar crater that is about to enter the umbral shadow. If you are organized, you will have a list of such craters ready ahead of time. A very useful book written just for this purpose, containing a catalog of 383 craters for lunar eclipse crater timings and a helpful map, is *Table and Schematic Chart of Selected Lunar Objects* by S.M. Kozik. I recommend you begin with the following craters on your "timings list", listed here in order (W to E) of their immersion in the umbral shadow:

1. Grimaldi
2. Aristarchus
3. Kepler
4. Copernicus
5. Pytheas
6. Timocharis
7. Tycho
8. Plato
9. Aristoteles
10. Eudoxus

11. Manilius

12. Menelaus

13. Plinius

14. Taruntius

15. Proclus

It obviously helps to know your lunar geography well, so that you can quickly locate the craters you want to time. Once you have located your target crater, use the techniques described later in this chapter for timing the occultation of bright stars by the Moon to record the exact instant at which the west rim of the crater enters the umbra, followed by similar timings for immersion into the umbra of the center and the east rim of the crater. Then record the respective times that the west rim, center and east rim emerge from the umbral shadow. Today, many advanced lunar eclipse observers prefer to use video recorders to get crater timings, as this method is more reliable than visual estimates and manual recording methods.

Time as many craters as possible for each total lunar eclipse – the more data you gather, the more valuable the observation for determining the precise geometry of the umbral shadow, which, as we have mentioned, is affected by the Earth's atmosphere, and which varies from one lunar eclipse to the next. Since we can calculate exactly how big the umbral shadow ought to be for each eclipse, simply from the geometry of the eclipse itself, we may subtract this value from the total size of the umbral shadow obtained from crater timings. The difference is the enhancement of the umbral shadow by Earth's atmosphere. No comprehensive model yet exists to accurately predict this enhancement factor.

Crater timings are designed to measure the precise size of the umbral shadow. But what of its intensity and spectral (color) characteristics? These factors are together known as the Danjon luminosity function, but predicting this quantity for any given lunar eclipse has proven as difficult as precisely predicting the umbral size. Table 11.1 gives

the Danjon scale, which can be used to estimate the intensity and spectrum of the umbral shadow. You may wish to indicate intermediate L-values by fractions or decimals, for example $L = 2\frac{1}{2}$ or 2.5. Make careful notes describing the visibility, tone, and color of the penumbral shadow leading up to and following the umbral phase, the color and tone of the umbral shadow, its edge, whether the edge of the umbral shadow was sharp or diffuse, the visibility of lunar surface features within the umbral shadow, and how each of the foregoing characteristics varied with time, if at all. Make careful notes of the time(s) you observed each of these characteristics, and record the local weather and sky conditions, as well as any factors related to the observing site that may come into play when recording your observations, for example thick smog.

Cloud and smog can obviously make the totally eclipsed Moon look much darker and/or dimmer than it normally would be in a clear, unpolluted sky – I have seen the eclipsed Moon completely vanish from view because of volcanic dust in the upper atmosphere! Sky conditions also have a profound effect on the color of the eclipsed Moon, which will affect L. If you have any artistic ability you will want to draw the eclipsed Moon, preferably in color. I have made many drawings of lunar eclipses, and it is interesting to see how varied the colors are. Of course, photography or video is superior for this purpose to drawings. Chapter 12 discusses how to photograph a lunar eclipse.

Besides the foregoing "qualitative" descriptions of lunar eclipse phenomena, it is possible for the amateur to undertake more quantitative studies, especially visual or photometric photometry of the eclipsed Moon or the brightness of specific lunar features during eclipses. Photometry just means measuring the brightness of something. We have already seen that during a total lunar eclipse the brightness of the Moon drops by 10 to 17 magnitudes, from –12.7 (uneclipsed) to –3 (very bright eclipse) or all the way to +4 (very dark eclipse). How can we tell how bright the totally eclipsed Moon

Table 11.1. The Danjon scale	
Danjon luminosity (L)	Qualitative characteristics of umbral shadow
0	Very dark eclipse. Almost invisible, especially at mid-eclipse.
1	Dark eclipse. Gray or brownish coloration, details discernible only with difficulty.
2	Deep red or rust-colored eclipse, with a very dark central umbra and the outer edge of the umbral shadow relatively light-colored.
3	Brick-red eclipse, usually with a bright or yellow rim around the umbral shadow.
4	Very bright, copper-red or orange eclipse, with a very bright bluish rim around the umbral shadow.

really is? We can't simply look at bright stars or planets in the sky around the eclipsed Moon, because the Moon is a huge, extended object with an angular size of $\frac{1}{2}°$, whereas stars and planets are point sources of light to our eyes.

To get round this problem, you may resort to a variety of tricks to reduce the apparent size of the Moon or to increase the apparent size of the star, or both. Viewing the Moon through reversed binoculars to make it appear fairly small may allow you to compare its brightness with the very brightest stars and planets. Other lunar eclipse observers have used common household devices such as hubcaps, silver reflective Christmas ornaments, or convex bicycle rear-view mirrors to reduce the Moon's angular size. You can also look at an out-of-focus star in your telescope to compare it to the eclipsed lunar disk.

The methods for using these devices are described in detail in the American Lunar Society's very comprehensive *Lunar Eclipse Handbook* by Francis G. Graham and John E. Westfall. Equations taking into account factors like atmospheric extinction are necessary when you are processing the visual photometric data – the *Lunar Eclipse Handbook* also explains these equations and how to use them. Some advanced lunar eclipse observers use a device known as a photoelectric photometer rather than their eyes to measure the brightness of either the full moon during a total eclipse or of specific lunar features such as maria or large craters. Francis Graham's photoelectric photometry setup is shown in Figure 11.3.

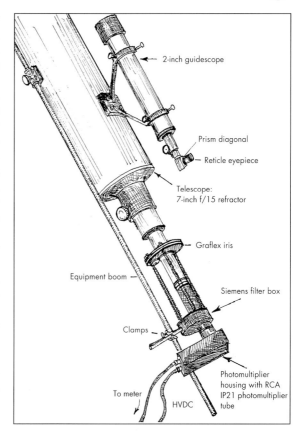

Figure 11.3. Francis Graham's photoelectric photometry equipment. *Illustration by Francis Graham.*

Lunar Occultations

Simply put, the Moon is said to occult a star when it interposes itself between the star and an observer on the Earth. They are fast-moving events – the Moon's orbital velocity is 3500 km/h (2200 mile/h), and it appears to move through 0″.5 per second of time. That means it takes the Moon about an hour to move its own width across the background stars, so no occultation can last longer than that period of time.

A *total occultation* occurs when any part of the Moon's disk apart from its extreme upper or lower limb completely covers the star. Every night the Moon totally occults many stars that happen to lie in its path along the ecliptic – these are hardly rare events, for the Moon's apparent diameter in our skies is 0″.5. Total occultations are fun to watch, especially when a very bright star suddenly vanishes behind the dark western limb of the Moon (remember that the Moon appears to move in an easterly direction against the sky's fixed background stars). In times past, astronomers used total lunar occultations to make accurately determinations of both the position of the Moon and the observer's latitude and longitude on Earth. Today these events are still useful for discovering new double stars, as they are capable of separating stellar components that are only 0″.02 apart!

More interesting, but much more challenging to observe, are so-called *grazing occultations*, or *grazes* for short. These occur when the Moon's northernmost or southernmost edge (limb) grazes a bright star, making it (barely!) disappear. A more scientific definition is provided by David Dunham, who wrote an important computer program while a graduate student at Yale University that allows us to predict grazing occultations: when a star is 3″ or less under the mean limb of the Moon at deepest occultation or multiple events caused by the star blinking on and off as it passes behind mountains on the lunar limb, we can define the event as a grazing occultation. As these events by definition occur at or near the

Moon's limbs, they are much shorter-lived than total occultations, often lasting only seconds instead of up to an hour.

Patrick Moore identified a third kind of lunar occultation, referred to as a *fading occultation* or *fade*. Some of these occultations are caused by the disappearance of first one component of a double or binary star system, then the other; others may be a consequence of refraction. The BAA has a "Project Fade" in force to study this phenomenon.

Because the limbs of the Moon are not the smooth edge of a perfect, polished sphere as imagined by ancient peoples, but display a jagged profile of alternating mountains and valleys, the occulted star may appear to blink on and off as these topographic features pass over it. The task of the occultation observer is to make highly accurate timings for each disappearance and reappearance of the star until it is completely clear of the Moon's limb. These timings provide valuable data related to the Moon's motion and topography. The first graze was probably observed near Santiago, Chile in 1852 by James Melville Gilliss of the US Naval Observatory.

Those wishing to observe grazing occultations must have, above else, a portable telescope. This is required because, unlike total occultations, grazes may be seen only from a very narrow path – often little more than 1 km (less than 1 mile) wide – on the face of the Earth. In fact, the most difficult part of observing a graze is to know exactly where you should be! There are ways of predicting where graze paths will lie, which I shall briefly discuss a little later in this chapter, but many occultation observers rely on carefully prepared maps that periodically appear in astronomical almanacs, the popular astronomy magazines, the handbooks of ALPO, the BAA, and the Royal Astronomical Society of Canada, and the newsletters of the ALPO and BAA Lunar Sections.

These sources usually try to provide predictions and maps for grazing occultations of all stars to at least magnitude 7.0, with predictions for stars as faint as 8.5 if the Moon is 40% or less sunlit (the brighter the Moon, the more difficult it is to observe fainter stars). The primary source for these predictions is the US Naval Observatory in Washington, DC. A dozen or so grazes may occur over the United Kingdom in a typical calendar year, but well over two hundred such events may be observed annually from North America.

But even these maps, prepared with the aid of computers, do not promise occultation observers success in catching their quarry – they only offer "predictions" for the paths of the events. Figure 11.4 shows a map which plots the paths of several lunar

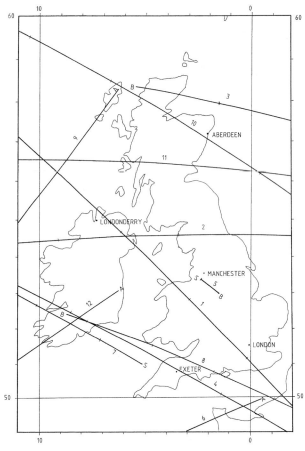

Figure 11.4. The paths of several lunar grazes predicted to occur over England during the year 2000. *Courtesy The British Astronomical Association.*

grazes predicted to occur over England for the year 2000. Therefore, if you have not figured it out already, besides a portable telescope, the most important requirement for successfully observing grazes is tenacity, for it is often frustrating to prepare for an occultation, only to discover at the moment of truth that you are slightly outside the narrow path. On the other hand, because it is so difficult to know just where to observe a graze, when you are successful it is a moment of immense joy and satisfaction!

It is preferable to use a clock-driven telescope for occultation work, as you will have your hands full (literally) just trying to accurately time the disappearance(s) and reappearance(s) of the target star. You don't want to worry about having to move your telescope by hand to keep the target star in the field of view, as not only will this distract you from the real job in front of you, but you will risk inaccurately moving the telescope so as to lose the target star at a critical moment. How horrible it would be to spend much time preparing to be in the precise path of a grazing occultation, only to miss the "main event" because you moved your telescope clumsily at the last second! For this reason, the now very popular Dobsonian telescopes are probably not the best choice for occultation work.

This places the prospective occultation observer in a bit of a quandary, both because many amateur astronomers acquire a Dobsonian as their first telescope, and because clock-driven, equatorially mounted instruments tend not to be very portable above 200 mm (8 inches) aperture. Even a telescope that size can be quite a lot to handle in the open field. The solution is to acquire a telescope of modest size for occultation work, if you don't already have one. Because most observable occultations are of fairly bright stars, large apertures are not needed.

In fact, the subtle dimming of a star passing barely behind a mountain peak on the Moon's limb is sometimes better seen in smaller telescopes, whereas observers equipped with larger apertures may not notice the effect, because of the greater light-gathering power of their instruments. Low-power eyepieces are usually best for occultation work, as they allow you to find the target star easily and keep it in the field of view with minimal effort. If the target star is faint or the Moon very brightly lit you may need to use a medium-power eyepiece, but high powers are to be avoided except by the most experienced observers.

When I first got into amateur astronomy in the early 1970s, most observers were equipped with either a 75 mm (3-inch) refractor or a 150 mm (6-inch) reflector. Larger apertures were rare. Occultation work was very popular back then, since it is well suited to modest-sized instruments. If your first telescope was in this size range, but you have since put it into storage as you have caught what amateur astronomers call "aperture fever," (meaning that you've acquired a larger telescope(s) for deep-sky observing or other projects), perhaps it is time you dusted off the trusty old instrument to go "grazing"!

Occultation enthusiasts often observe in groups, which has its advantages. For example, one of you will need to be equipped with a short-wave or other kind of radio capable of receiving standard time signals, to calibrate your data. In the USA, the National Bureau of Standards operates short-wave radio station WWV that does nothing but broadcast time signals at frequencies of 5, 10, 15, and 20 MHz; MSF in Rugby, England does the same thing. Not very interesting radio programming, but indispens-

able to occultation "chasers." Occultations are all about timing – if you use standard time signals, the accuracy of your timings will be limited mainly by your "human reaction time", usually 0.3 to 0.5 s. That is how long it takes the typical experienced observer to react to the disappearance or reappearance of an occulted star and to voice-record the event, usually by speaking the words "out" or "in" into a recorder. "Out" simply means "out of sight" – the star has disappeared behind the Moon's limb; "in" means it has reappeared from behind the limb.

Thus the third piece of equipment you will need is a voice recorder, either an old-style cassette tape recorder or the more modern hand-held variety. Make sure well before the occultation that your setup allows the standard time signals to be easily heard on the audio tape. And, while this may sound a bit obvious, make sure your recorder has fresh batteries and bring extras just in case – more than a few observers of occultations have returned home after what they thought was a successful graze observation only to discover that their recorder wasn't working properly.

You should also keep a stopwatch as backup for the recorder, preferably the newer electronic "LED" display kind, which are usually accurate to 0.001 s. If you are using an old-style stopwatch with moving parts, make sure you can read its time accurately to at least 0.1 s, preferably 0.05 s. Test the stopwatch ahead of time for running accuracy – if it loses or gains more than 0.1 s per 10 min of running time, don't use it! Alternatively, you could use a mobile phone to call the telephone time service.

You will also need a good topographic map to determine your exact longitude and latitude (to within 15 m or 50 ft) and elevation above sea level to the nearest 30 m (100 ft). Unless you are observing the occultation as part of a team led by an experienced observer, you will also need the topographic map to get to a location on or near the predicted path. If you have a Global Positioning System handset, this will give an accurate fix of your position.

In recent years, many "graze-chasers" have begun using video cameras like the ones described in the following chapter to videotape occultation events. Obviously, this method has two big advantages over the visual method – your reaction time does not affect the recording of the precise moments of disappearance and reappearance of the target star, the video camera instantly capturing on film the instants of these events, and the video camera can image fainter occultations than the human eye can detect. The video recorder will also stand a better chance than you of recording multiple subtle dimmings of the target star. Most lunar occultation experts now agree that to further advance our knowledge of the lunar limb profile or the Moon's orbit through grazing occultations, video techniques will have to be used, as we have gone about as far as we can go with purely visual recordings of these events. As with the tape recorder, make sure your camcorder's battery is charged!

If the clock on the video recorder is accurate enough, which is a big "if," you may also eliminate data reduction – interpolating between the standard time signals to get exact times for disappearance and reappearance. But I wouldn't count on this. The video method is an update of the "clock-and-camera" method whereby observers would set up a camera aimed at an accurate, easily read clock, then take photographs of it at the instants of disappearance and reappearance of the target star. Some occultation observers still prefer this method, which has the advantage that nothing has to be written down.

If you don't possess a video camera (and maybe even if you do!), you should still try your hand at visual occultation work, as there are few better introductions to "serious" amateur astronomy. The skills that you develop as an occultation observer are applicable to many other observing exercises, and the experience gained will serve you well. Occultations are also a lot of fun and a great way of getting groups of people to take part in organized observing projects. Because grazing occultation paths are so uncertain, it is standard practice to space observers out along a line perpendicular to the predicted path, so that at least one or two of the observers will witness the event(s) – the more the merrier!

Beginning occultation observers should get a few total occultations under their belts before trying to observe their first graze. This is to get a feel for the equipment and timing procedures, to get used to the speed of the Moon's against the background stars, and, most of all, to get used to the instantaneous disappearance of the target star, which is guaranteed to take you by surprise at your first occultation. If the Moon had an appreciable gaseous atmosphere to scatter and refract the target star's light, the disappearance would be more gradual, but without such an atmosphere the star vanishes with startling suddenness.

You should time the disappearance and reappearance of the target star to the nearest 0.1 s. Astronomical almanacs predict total occultations for 18 so-called standard stations all over North America,

and for Greenwich, England, and Edinburgh, Scotland. There are many other standard stations throughout the world, including Sydney and Melbourne in Australia and Wellington and Dunedin in New Zealand. The most valuable data for total occultations are obtained when the target star passes behind the Moon very near its equator, as this gives the greatest elapsed time for the event and the smallest errors in predicted times.

Once you have gained proficiency in observing total occultations – you may want to try observing these events in the company of more experienced observers who can make sure you are on the right track – you are ready for your first graze. If you can find a group of observers organized in your geographic region to go on graze expeditions, by all means take advantage of their expertise and ask to join them on their next outing, as your chance of successfully witnessing the graze events will thus be greatly increased. While it is generally advised to observe far enough away from other members of a graze team so that the two of you do not interfere, verbally or psychologically (if one observer hears another recording events, that might influence their own observations) with one another, for your first graze you may wish to quietly observe next to a more experienced observer, to check your own observing skills and methods.

One word of caution is in order. Since observing occultations requires you to be in a very precise location along a very narrow path across the Earth's surface, you will frequently find yourself having to set up your telescope on someone else's land. Ask permission first, of course, explaining that you are an amateur astronomer who is trying to get data to submit to internationally recognized scientific organizations. If you observe as part of a team, presumably the team leader will have obtained this permission long before you arrive at the observing site.

All occultation reports on standardized forms should be submitted to one of the organizations listed in "Resources for the Lunar Observer" (on the CD-ROM). The principal groups that issue predictions and collect and utilize data from occultations are the International Lunar Occultation Center in Japan, the International Occultation Timing Association (IOTA) in Topeka, Kansas, and the BAA's Occultation Subsection (part of the Lunar Section). These organizations offer handbooks, newsletters, and other materials to help observers interested in contributing data to them.

Chapter 12

Imaging the Moon

When I first started observing the Moon in the early 1970s, it didn't occur to me that I could take pictures of this world that astronauts were exploring at that very time. I assumed that only NASA, with its multi-billion dollar budget, could do such wonderful things. I was therefore excited to learn, a few years later when I joined an amateur astronomy club, that "regular" people could take good-quality photographs of the Moon using home-made telescopes and ordinary 35 mm single lens reflex (SLR) cameras. I still recall the excitement of taking my first lunar photographs, through a classic Cave 405 mm (16-inch) telescope that belonged (and still does) to the Syracuse Astronomical Society in upstate New York. They weren't very good, but no matter – I was happy that I got any images at all!

Little could I, or other amateurs at that time (the late 1970s), imagine what advances in technology lay ahead for imaging the Moon and other astronomical objects, nor would we have believed that such technologies would someday become available, not just to professionals in their mountaintop observatories, but to amateur observers as well. But that is exactly what happened in the past twenty years. Today, amateur astronomers who wish to image the Moon have two options available to them that simply did not exist only a couple of decades ago – video and CCD imaging.

These new technologies have opened up a whole new world of possibilities for the amateur lunar enthusiast. The results obtained by some of the more talented amateurs using these imaging technologies are of professional quality – as you will see from some of the examples presented in this chapter. Many amateurs still prefer more traditional astrophotography of the Moon with regular cameras and film But some of them have applied this older technology to new and exciting projects, so we shall

also look at a few examples of this kind of work. Because I still take horrible lunar photos, I will step aside for this, the last chapter of the book, and let those who are more talented than me share their secrets with you. I shall begin with imaging the Moon using ordinary photography.

Many excellent books have been already been written instructing the amateur astronomer in the use of an ordinary 35 mm SLR camera to obtain quality astrophotographs of the Moon, so our discussion of this topic will be fairly brief. One of the books in this series, *Astronomical Equipment for Amateurs* by Martin Mobberly, covers astrophotography with a thoroughness that is admirable. Besides this excellent book, I would strongly recommend the work by Tom Dobbins, Donald Parker, and the late Charles "Chick" Capen titled *Introduction to Observing and Photographing the Solar System*, which is still in print. Instead of rehashing what other guidebooks have to say on the subject of astrophotography, I thought I would take a fresh approach by describing a couple of interesting applications of photography to amateur lunar studies.

There are basically two ways in which you can use a 35 mm SLR camera to image the Moon – you can take photos through a telephoto lens, or you can attach your camera to a telescope. Amateurs frequently overlook the advantages of telephoto lens work with the Moon, which I think is a mistake. English lunar imager Michael Oates has a wonderful Website (see "Resources for the Lunar Observer" on the CD-ROM) which serves as a superb "how-to" guide for taking photos of the Moon with telephoto lenses. Here he passes on some of his helpful advice.

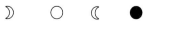

Photographing the Moon with Telephoto Lenses

Michael Oates

The techniques I describe here work well for me, but I have no doubt that there are other ways of getting good results, so if you find any way of taking photos of the Moon that you are pleased with, then stick with it and enjoy. Don't go mad and buy the most expensive equipment, just make the most of what you have. High-quality work can be done with inexpensive equipment – it's how you use it that matters. I shall be discussing lunar photography through camera lenses only, but a lot of the techniques are also applicable to lunar photography through a telescope.

Lenses

One big mistake beginners make when trying to photograph the Moon is using a lens with too short a focal length. To the naked eye the Moon may seem quite large, with quite a lot of detail. However, when photographed, the resulting image is often far too small and looks like a small dot on the print or slide, with little or no detail recorded. In order to show how large (or small) the Moon appears on a 35 mm film frame, I have taken a number of photographs of the Moon with different lenses ranging from a wide-angle lens of 24 mm focal length right up to a 1600 mm focal length telephoto (an 800 mm lens with a ×2 tele-converter – see below), which you can view on my Website. Table 12.1 gives the size of the Moon's image on 35 mm film when taken with various focal length lenses. You really need a focal length of 800 mm or more, the best being 1600 mm, which allows the Moon to fit in the frame easily. You can use the following formula to determine the image size on 35 mm:

Focal length of lens (mm)/110 = Size of image (mm)

Tele-converters

A tele-converter ("Barlow lens") is a special lens that typically doubles or triples the effective focal length of the camera system, allowing you to enlarge the size of the image of the object being photographed. It is often said that tele-converters degrade

Table 12.1. The size of the Moon's image on 35 mm film frames (all values in millimeters)

Focal length	Size of image
24	0.22
28	0.25
50	0.45
80	0.73
135	1.23
200	1.82
300	2.73
500	4.55
800	7.27
1000	9.09
1600	14.55

the image and should be avoided, and that they also lose a lot of light, typically two stops for a ×2 converter. I have found by experience that an image produced with a tele-converter is *far superior* to an image taken with the same lens without the converter and having the image enlarged to the same scale. Why? Well, it's to do with the resolving power of the film – if you enlarge a negative or transparency, you enlarge the grain, and the amount that the tele-converter degrades the image is less than what you lose by enlarging the film. And you have the added advantage of having a larger image in the first place. I cannot say this of ×3 converters, as I have only used a ×2, and I would certainly recommend only using good quality multi-element converters that have all their optical surfaces coated.

You need not stop at one tele-converter: I regularly use two ×2 converters together, and the results are fantastic – far better than I would get by enlarging the image afterwards. As for light loss, we're talking about photographing the second brightest object in the sky, so we can afford to sacrifice a bit in order to get better images. I would like to add that a 1000 mm lens is likely to produce better images than a 500 mm with a ×2 converter, so if you have a choice, use the longer lens.

Cameras

Most of my photographs of the Moon were taken with a Praktica MTL5B, a 35 mm manual SLR camera that takes low-cost 42 mm thread lenses. The camera was also very low-cost. When the shutter release button is depressed, two things happen in an SLR: the mirror is moved out of the way to take the picture, and the shutter opens. It is essential that very little vibration occurs at this time. Any vibration from the mirror or shutter will cause the image to blur, and the blurring will be more pronounced the longer the focal length of the lens. The camera must have a cable shutter release, to reduce vibrations that may be caused by operating the shutter manually. There are ways of stopping the vibration caused by the mirror moving. One is to lock the mirror out of the way before the exposure, but this has the drawback that you can't see the Moon in the viewfinder just before the exposure, which could result in the Moon's image being off-center or even off the image altogether. Another method is to hold a piece of black card in front of the lens, making sure you don't touch the lens; press the shutter using a cable release, wait a few seconds for the vibrations to die down, then quickly move the card away. Replace it to end the exposure, and with the card still in front of the lens, close the shutter.

An automatic camera is not necessary, indeed it could be undesirable as underexposure could result if the light sensor of the camera is in the center of the image, where the bright Moon will cause the camera either to stop down the lens or to reduce the exposure time. The opposite may happen if the sensors pick up more of the background. Also, automatic cameras tend not to work very well in the sub-zero temperatures in which we astrophographers sometimes operate. It's better to use a manual camera; if you do use an automatic one make sure that it has a manual override. A mechanical shutter is preferable to an electronic shutter, which can experience problems at cold temperatures. I shall deal with how to calculate the exposure below.

Mounting

For most lunar photography, if you are not using lenses of very long focal length, or if the exposure is shorter than about half a second, then a sturdy tripod is all that's needed. But for long focal lengths and longer exposures a driven mount is required. This can be a small mount that fits onto a camera tripod, a larger equatorial mounting. I can't emphasize too strongly the necessity of a good strong tripod that does not vibrate like a tuning folk every time the shutter is pressed. If you can't keep the camera still during an exposure you will be very disappointed in the results.

Film

As the Moon is a bright object, a slow fine-grain film can be used – indeed, the best results will be obtained with the slowest films: Kodak Technical Pan 2415 for black and white photographs, Kodachrome 25 for color slides. Both these are ISO 25 rated films and are capable of producing very sharp, well-defined images. As you move to faster films, the film grain starts to degrade the image. There are times when faster films are an advantage: when faster shutter speeds are needed, if a motorized camera mount is not available, or when photographing a lunar eclipse, when much longer exposures are needed.

Exposure

The easiest way of calculating the exposure is to use the Exposure Calculator available at my Website. With this calculator the exposure can be determined for any speed of film and lunar phase. You should *bracket* the exposures, i.e. you should take three exposures, one overexposed by one or two stops, the second exposure as indicated, and the third underexposed by one or two stops. The reason for this is that so many variables affect the correct exposure time, that it is prudent to allow yourself a wide margin for error. Table 12.2 gives the exposure time, in seconds or fractions of a second, for ISO 25 film.

Here are three different methods of actually taking lunar photographs. Choose the one that best matches your situation.

Long focal length lens on camera drive

I shall assume that the lens used is an 800 mm *f*/8, film used is ISO 25, and the Moon is at quarter phase. Roughly polar-align the camera mount, and get the Moon's disk in the center of the viewfinder. If the motor drive is switched on, it will keep the Moon in the center of the viewfinder for quite a while. Focus the image and set the exposure on the camera – in this case I am assuming an exposure of 1 second at *f*/32, which was obtained from the Exposure Calculator on my Website.

When you are ready to take an exposure, use the black-card technique described above. I would recommend repeating the same exposure a few times in the hope that one is sharp. You should also take another exposure at slightly shorter time and one at a slightly longer time, to increase the chance of making the right exposure.

Table 12.2.   Exposure times, in seconds, for ISO 25 film for different lunar phases

f-stop	Thin crescent	Wide crescent	Quarter	Gibbous	Full
1.4	1/125	1/250	1/500	1/1000	—
2	1/60	1/125	1/250	1/500	1/2000
2.8	1/30	1/60	1/125	1/250	1/1000
4	1/15	1/30	1/60	1/125	1/500
5.6	1/8	1/15	1/30	1/60	1/250
8	1/4	1/8	1/15	1/30	1/125
11	1/2	1/4	1/8	1/15	1/60
16	1	1/2	1/4	1/8	1/30
22	2	1	1/2	1/4	1/15
32	5	2	1	1/2	1/8
45	15	5	2	1	1/4
64	45	15	5	2	1/2
90	—	45	15	5	1
128	—	—	45	15	2
250	—	—	—	45	5

Long focal length lens on fixed tripod

With this setup you have to make a choice: you can use a fast film, and a high shutter speed, or a slower film and the black card technique, as above. But in either case you will be limited to exposures of less than 1 second, preferably less than 1/2 second, because for longer exposures the Moon will move too much, causing the image to blur.

Short focal length lens on fixed tripod

If shorter focal length lenses are used, any movement of the Moon in short exposures will not degrade the image as there will be little detail recorded anyway. Also, the use of black card for the exposure may not be required (it will have to be determined by experiment). To record as much detail in the lunar features as possible, a fine-grain film should be used, otherwise any detail will be lost in the grain. If longer exposures of up to 20–30 seconds are taken with standard lenses, using faster film (ISO 400 or higher) and wide apertures, you will record some of the brighter stars and planets (if they are in the same area of the sky). These photos can be very pleasing. At conjunctions, try to record as many Solar System bodies as possible in one frame (note that the Moon will be overexposed).

Tips

Don't be afraid of taking too many photographs in one session. Shoot a whole roll of film if you can afford to do so, for this will increase your chances of getting one or two good shots. There are a great many obstacles in our way of getting that perfectly sharp image:

- camera shake through use of poor tripod,
- camera mirror and shutter induced vibration,
- wind,
- poor seeing (turbulence in the atmosphere).

Use a lens hood to stop stray light from street lights, etc. from striking the lens. Stray light causes loss of contrast and internal reflections, producing flares on the image. It will also go a long way to preventing the lens from dewing up. Unless you can process and print your own photographs, use slide film. You will be very fortunate if you know of a commercial photo-lab which can make good prints of astronomical images, even images of the Moon.

Lunar Eclipses

Taking photographs of lunar eclipses requires a few different techniques. The Moon is normally a very bright object, but during a total eclipse the Moon becomes very dark, exposures need to be longer and faster films may also be needed. The actual brightness of the Moon during an eclipse varies a great deal, not only during a single eclipse, but also from one eclipse to another. Weather and the amount of dust in the atmosphere can cause the Moon to almost disappear from view. For these reasons, to give precise exposures for an eclipse is not possible – one can give only general guidelines. You should take several exposures with a wide range of exposure times to ensure that you get some well-exposed pictures.

The Moon does not suddenly disappear during an eclipse – it first enters a penumbral phase where the Moon's disk is only partially shaded. The penumbra moves across the disk, followed by the umbral shadow. In order to try and capture on film the advancing edge of the shadow, you might need to underexpose slightly; overexposing will just burn out the image, and the shadow may not be seen. As longer exposures are needed, you may need to drive the camera to keep the Moon's image sharp. Alternatively, you can use a lens of shorter focal length, but this will make the image much smaller. Faster films will help to keep the exposure length down, but the faster the film, the more obtrusive the grain size. Color film is better than black and white, to record the lovely reddish color the Moon takes on during totality. You may also be able to record stars very close to the Moon, which is something that's not normally possible because the uneclipsed Moon is so bright.

Not all eclipses are total – many are partial, where the Earth's shadow does not completely cover the lunar surface. There are also penumbral eclipses where the Moon enters the only the penumbra of the shadow. It can be quite a challenge to get a good image of a penumbral eclipse, as there is no clearly defined shadow, the difference in brightness from the normal appearance being very slight. A good starting point for exposure when photographing the penumbra is 1/60 second at *f*/8, using ISO 100 film. This is the same exposure as used when photographing the uneclipsed Moon at gibbous phase (use the Exposure Calculator). For exposures of the umbral phase, to record the shadow's edge with the lighter part of the Moon exposed correctly, try 5 seconds at *f*/8 on ISO 100 film (this is only a very approximate suggestion).

More advanced imaging projects are possible if you combine your 35 mm SLR camera with a telescope, so that the telescope becomes, for all practical pur-

poses, a giant telephoto lens. With the increased resolving power of the telescope, your film will now capture details that are not possible with ordinary telephoto lenses. Much shorter exposures, which allow you to obtain much sharper images, are now possible because of the telescope's superior light-gathering ability. But what kinds of project can you undertake? Here are two very different, imaginative projects undertaken by Bill Dembowski, who is active with the American Lunar Society and who serves as Topographic Coordinator for ALPO's Lunar Section. One of these projects uses photography to make lunar height measurements.

Lunar Photography: An Apogee–Perigee Project

Bill Dembowski

Part of the appeal of photography is its ability to record lunar detail with unquestionable accuracy. While it's true that no photograph taken through a telescope can match the wealth of detail visible to the human eye through the same telescope, no sketch can match a photograph for the accuracy of what it does record. A good example of photographic accuracy is a project I undertook in 1992 to illustrate the difference in the apparent size of the full moon at apogee and perigee. The Moon travels about the Earth in an elliptical orbit, so its distance from the Earth is not constant. At its greatest distance (apogee) it will appear to be smaller that it does at its nearest point (perigee). The difference is not apparent to the eye, but should be noticeable in a series of photographs.

The first step was to determine when apogee and perigee would be the most extreme (not all are equidistant) and occur at full moon. A quick look at Guy Ottewell's *Astronomical Calendar* showed that the full moon in January would be 363 263 km (225 727 miles) from Earth, while the full moon in August would be a whopping 405 547 km (251 995 miles) distant. My target dates were, therefore, January 19 and August 13.

In a project of this sort, where comparisons are to be made between two different photographs, one must be certain to take both photos with the same equipment in the same configuration. I had to be sure that I used not only the same telescope, but that all other parts of the setup were the same as well. I have, for instance, several different adapters for linking a camera to my telescope. If I used two different adapters, having slightly different lengths, I would negate the project by introducing a variation in size that was not the result of the Moon's orbital position. I decided to use a 250 mm (10-inch) Schmidt–Cassegrain telescope operating at *f*/5 since this would give me the approximate image size I wanted.

January 19, 1992, the night of perigee, was the coldest night of the winter. Working quickly to beat the cold and some rapidly approaching bad weather, I managed to get several usable shots. Although the presence of some thin, high cloud caused some loss of contrast, the results were still acceptable. Photographing the August Moon was almost leisurely by comparison, though predictions of rain had me worried for days in advance.

The next critical step in the process was making the prints. Processing film cannot alter the size of the final image, but making the enlargements most certainly can. The ideal method is to place both negatives in the enlarger at the same time. This will guarantee that both images are being enlarged to the same degree. Depending on the actual size and position of the images, however, this method may not be possible or aesthetically pleasing. Another way would be to project first one image then the other onto the enlarging paper without adjusting the height of the enlarger. Or I could simply make two separate prints using the same enlarger height. I felt that the images would have more impact if both were on the same print, so I sandwiched the negatives and printed them both at the same time.

The resulting photograph (see Fig. 12.1) clearly shows the difference in apparent size. I suppose that, using these two images as a baseline, it would now be possible to determine the Moon's distance at any given time by carefully measuring its diameter on any subsequent negatives taken with the same equipment. The important thing, however, is to look at lunar photography in new and creative ways in order to more fully utilize its ability to record with great accuracy, and to enhance the enjoyment of lunar astronomy.

Vertical Studies

Bill Dembowski

One of the most rewarding pursuits for the amateur engaged in lunar observing is the determination of the heights of lunar features. Two goals of amateur measurements of vertical lunar distances are to redetermine heights of the major features to check their accuracy, and to determine the heights or depths of lunar features for which vertical relief has not been previously measured. Once characterized by tedious mathematical calculations, this noble pursuit is now made infinitely more simple and enjoyable by Harry Jamieson's collection of software programs, the *Lunar Observer's Tool Kit*. When using the *Tool Kit* one need only furnish certain basic information, and the program does the number crunching for you. The data required are as follows:

Figure 12.1.
The differences in apparent size of the Moon at apogee (left) and perigee (right). *Photographs by Bill Dembowski.*

1. The *location of the observer* in degrees and minutes, and the elevation in meters (to the nearest 10 m). The best source for this information in the USA is a geodetic map published by the US Geological Survey, often available in sporting goods stores and/or public libraries. In the UK, refer to an Ordnance Survey map.

2. The *date* and *Universal Time*, to the nearest minute.

3. The *location of the lunar feature* being measured, in degrees and minutes of latitude and longitude. The better the lunar atlas used, the more accurate the results. My personal choice is the *Orthographic Atlas of the Moon*, which allows interpolation of positions down to 1.6 km (1 mile).

4. The *length of the shadow* cast by the feature. This information may be in kilometers, fractions of the lunar radii, or seconds of arc. Obviously, accuracy is critical here. Several methods of shadow measurement are available to the amateur, the most common being:

   (a) *Estimating by eye.* This method is the least accurate and is performed by visually comparing the length of the shadow with a feature of known size. If the shadow length appears to be half the size of a 30 km crater, the value is entered as 15 km.

   (b) *Drift method.* Here the shadow is allowed to drift across the reticle (cross-wire) of a high-power eyepiece, with the clock drive running, while timing its passage. That time is then compared with the time it takes for a feature of known size to drift across the same reticle. As a rule of thumb, the Moon will drift from west to east at the rate of slightly less than 1 km s^{-1}.

   (c) *Bifilar micrometer.* Probably the most precise method of shadow measurement is achieved with the aid of a bifilar micrometer. Measurements can be made by comparing the length of the shadow with the size of a known feature, or by calibrating the micrometer on a double star of known separation and measuring the shadow in seconds of arc.

   (d) *Graduated Reticle.* Eyepieces with graduated reticles, such as Celestron's Micro-Guide, may be used in the same manner as a bifilar micrometer. Although not as precise, they are less expensive and easier to use.

   (e) *Photographic.* Measuring shadow lengths on a photograph is similar in procedure to using a graduated reticle eyepiece. Results are comparable, but a photograph has the advantage of containing many shadows which can be measured under more controlled conditions, and the measurements are infinitely repeatable. This is the method I use, measuring the shadows with an architect's rule calibrated to 1/60 inch. These measurements are then compared with measurements of a fixed double star, photographed and printed under identical circumstances, to obtain the length of the shadow in seconds of arc.

Figure 12.2 is a photograph of the Eratosthenes region, with Sinus Aestuum at center and Mare Imbrium to the right. The Apennines are the moun-

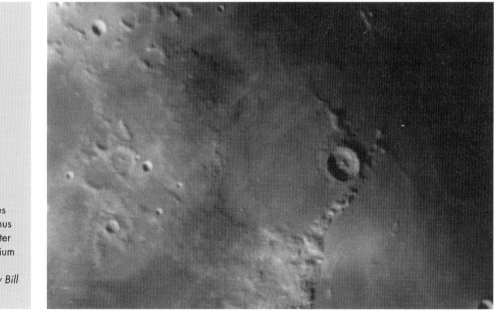

Figure 12.2.
The Eratosthenes region, with Sinus Aestuum at center and Mare Imbrium to the right.
Photographs by Bill Dembowski.

tains at the bottom of the photograph, which shows shadows that are easily measured as described above to determine the vertical relief.

The next section was written by an amateur astronomer who has developed a proficiency in the use of a video camera designed especially for use in astronomy, called the Astrovid 2000. Before this camera was developed, a few amateurs imaged the Moon by mounting ordinary camcorders on their telescopes, with occasional good results. But it is preferable to use an astro-video camera specifically adapted for use at the telescope, and these devices are now widely available, and affordable. As with CCDs, it is possible to process the images obtained by these cameras, which are particularly adept at imaging the Moon and bright planets. From this section you will get a good idea of how astro-video cameras work and how they can be used to obtain impressive images of lunar features.

Imaging the Moon with an Astro-Video Camera

Joe LaVigne

Four decades of lunar videography

Amateur astronomers have been imaging the Moon with video cameras since the 1960s. Video pioneers started by mounting large closed-circuit cameras to their telescopes so that they could view the Moon and planets on TV monitors, perhaps even sharing these views with large groups of people.

These early attempts were a bit awkward as the cameras were bulky, and were never designed to be mated to an astronomical telescope. The astronomer had to shoot video through the camera lens and through the eyepiece of the telescope itself, called the "afocal" method. This meant that the alignment between telescope and camera lens had to be perfect or the image would be blurred. Since

these cameras were heavy, they may have had to be mounted on a tripod, which made tracking of the astronomical object difficult if not impossible. If an amateur were to mount the camera on the telescope itself, then brackets had to be constructed and the camera counterweighted in order for the telescope to track correctly.

In order to videotape what they saw on the monitor, these pioneers had to possess a videotape recorder of the day, which was generally a bulky reel-to-reel device. This made sharing their results with others somewhat difficult. Suffice it to say that astro-videography wasn't widely practiced by amateurs in the early days because the equipment needed was not readily available to the average person.

This all began to change in the late 1980s with the advent of the hand-held video cameras that flooded the consumer electronics market. These cameras solved several of the problems encountered by early videographers. The cameras were readily available, they were light, their prices were dropping, and – most importantly – they recorded onto videocassettes. The afocal method was still necessary because the lenses of these cameras were not removable. With their lighter weight, camera brackets and counterweights were not too difficult for a handy amateur to fabricate. So now lunar and planetary videography was possible, and it was easier to share results with fellow amateurs.

With the advent of camcorders came the idea of using camcorder components to make dedicated astronomical imaging cameras. In the late 1980s several companies began to use the CCD chips from camcorders in small lightweight cameras designed specifically to mount into the focuser of a telescope. By eliminating the lens and the videotape components of a camcorder, they were able to build small cameras which contained just the circuitry and the CCD chip. These cameras could output either directly to a monitor or to a monitor/VCR combination. And because they were lightweight, they eliminated the need for awkward brackets and large amounts of counterweighting, which eased the tracking burden on the telescope.

Another technological advance that made astro-videography more appealing was the development of the personal computer. Computers became more powerful at the beginning of the 1990s. Using video-capture cards in their PCs, videographers were now able to capture single frames of video directly from these cameras or from video recordings.

No longer were enthusiasts restricted to simply editing and displaying videos of their imagery. Now they could capture an image, manipulate it to some

Figure 12.3. The crater chain Theophilus, Cyrillus, and Catharina taken with a 250 mm (10-inch) Newtonian telescope at *f*/12 using an Astrovid 2000 video camera. *Video image by Joe LaVigne.*

Figure 12.4. The Astrovid 2000 video camera and its control box, along with a Snappy video capture device. *Video image by Joe LaVigne.*

degree, and perhaps even print it out. With the further advent of image processing programs such as Adobe *Photoshop*, astro-videographers could now extract data from an image that was not readily apparent in the original. This allowed advanced amateurs to begin to do some real science with their images, extracting detail that could be useful to both amateur and professional astronomers alike. Figure 12.3 shows the crater chain Theophilus, Cyrillus, and Catharina taken with a 250 mm (10-inch) Newtonian telescope at *f*/12 using an Astrovid 2000.

What is an astro-video camera?

The term "astro" may be a bit misleading here. These cameras are built around commonly available camcorder chips and they are sensitive to low light levels, generally 0.05 to 0.01 lx. This makes them suitable for the brighter astronomical objects. Coupled to a telescope, they are easily capable of imaging the five naked-eye planets, as well as the Sun and Moon, but they are not necessarily sensitive enough to effectively image stars and nonstellar objects such as nebulae and galaxies. Most astro-video cameras on the market are black and white devices, so don't expect colored images of the Sun and planets. For lunar observation this hardly matters, unless you intend to study lunar transient phenomena (LTPs), for which color may be required.

The basic astro-video camera is essentially a lensless camera consisting of the basic CCD chip and its immediate support circuitry. The chip and circuitry come in a small metal case, 40–50 mm ($1\frac{1}{2}$–2 inches) square and 75–100 mm (3–4 inches) in length. These packages weigh approximately a 250 g (8 oz), depending on the brand. The chip sits inside a recessed circular aperture where it can be exposed directly to the telescope's focal point. A "C" adapter can be screwed into this aperture, allowing the camera to be inserted into a standard 32 mm ($1\frac{1}{4}$-inch) telescope focuser. The camera essentially replaces the eyepiece, and weighs no more than some 50 mm (2-inch) eyepieces. Figure 12.4 shows the Astrovid 2000 video camera and its control box, along with a Snappy video capture device.

A video output cable leads from the camera either to control box or directly to the monitor equipment. Video control boxes extend the capabilities of the camera, allowing the operator to adjust the exposure and contrast as well as increase the gain, which allows fainter features to be imaged. Some basic units have many of these parameters fixed, so that the user cannot adjust them.

The chips in astro-video cameras generally come in two sizes, the most common of which is called a $\frac{1}{3}$-inch chip, and there is larger chip called a $\frac{1}{2}$-inch chip; their actual sizes are a bit smaller. The chips are protected by an optical cover to prevent them from scratching and dust contamination. They come in two varieties – the interline frame transfer chip and the interline transfer chip. With the interline frame transfer chip, photons activate all the pixels on the chip, the data are then sent as a whole to the circuitry, and the chip is refreshed. This allows full use of the surface area of the chip. In interline transfer chips some pixels gather photons and then send a signal to an adjacent pixel, the data from these storage pixels are then sent to the circuitry, and the chip is refreshed. This type of chip doesn't allow full use of the entire surface area. While the interline frame transfer chip provides higher resolution, this doesn't mean that cameras using the interline transfer chip can't provide stunning images.

Like a camcorder, the astro-video camera is constantly taking images at a frame rate of 30 frames per second, refreshing the chip each time the data transfer is completed. This means that it is constantly updating what the telescope is seeing, and this can be very useful to a lunar imager.

Why video as opposed to film or CCD?

Video offers several advantages over traditional 35 mm photography and CCD imaging. With film the observer needs to work hard to achieve a fine focus. If the photographer fails at this task, then one or more shots are wasted and they won't even know it until all the darkroom work is done or the prints are returned from the photo lab. CCD cameras offer a similar challenge to achieve fine focus. The CCD imager needs to take focusing frames to ensure that the camera is in focus, then take the shot, check the focus, and repeat the process, perhaps several times. The CCD imager has the advantage of eliminating the lengthy darkroom process, but trial and error is sometimes required to get that good image.

Since both types of cameras require critical focusing before exposures can be made, they are also at the mercy of another factor – atmospheric seeing. After working diligently to achieve focus, the observer opens the shutter or presses the "Enter" key, not knowing if the sky will boil between the Moon and the camera in that critical fraction of a second of exposing the film or chip to moonlight. Video cameras are not as susceptible to the vagaries of focus and seeing. The image from the camera is displayed on a video monitor in real time. This means that if the wind shakes the telescope, the image moves at that very instant. If a cloud, airplane, or bird passes between Moon and telescope, the observer sees it as it happens. The observer can adjust the focus instantly in reaction to changing conditions. This eliminates the trial and error of sample focusing frames or dim viewfinders.

A typical astro-video camera takes 30 exposures (frames) per second, refreshing itself after each exposure. The exposure time of each frame can be predetermined, ranging from 1/60 s to as short as 1/100 000 s, depending on the camera control units. With so many short exposures per second being displayed continuously on the video monitor, the human eye is fooled into thinking that the seeing is steadier than it may appear visually through the eyepiece of the telescope. If the video output has been sent to a VCR, the observer has literally thousands of frames to choose from when it the time comes to frame-grab and image-process. This allows the videographer to discard scores of frames ruined by bad seeing, and choose one that has recorded good seeing.

One of the most remarkable features of a video system is its suitability for use in public demonstrations. Since the image is piped to a monitor or a TV, many people can view it simultaneously. With the image moving with the movements of the telescope, people are invariably awed by what they are seeing: when the operator slews the telescope, the audience sees the Moon "move" in real time. By extending the cabling, the image can be sent indoors to an auditorium where an audience can view the image in comfort. The output could even be routed to a video camera for direct use by local television broadcasters.

Cost

With all the new gadgets coming onto the market these past few years, many amateur astronomers are anxious to jump in and begin taking fantastic pictures of the Moon and planets, as well as deep-sky objects. For years the only choice was the 35 mm SLR camera, coupled to an equatorially mounted and clock-driven telescope. Crucial polar alignment was essential, and the better your equipment and techniques, the better your final output.

On the scale of imaging devices, 35 mm SLRs are relatively inexpensive, but unfortunately the same cannot be said for the support equipment required to obtain good results. CCD cameras are at the opposite end of the price scale, the camera units themselves can range from about three time the cost of a basic SLR at the low end to well over fifteen times as much for a premium unit. Add the same cost for a motor-driven telescope, as in the case of the 35 mm SLR, and then add the cost of a personal computer to complete the package. Fortunately, a large number of households in the USA and Europe already own personal computers; but not everyone is willing to pull their workhorse PC out into the night air to do CCD imaging. So in many instances a second PC or a laptop may have to be purchased.

Astro-video cameras come into the mid-price range for imaging devices available to the amateur. They provide more versatility than 35 mm photography as well as many of the advantages of CCD imaging. Basic $\frac{1}{3}$-inch video cameras can be had for little more than the price of a basic SLR. Premium units go for around the price of a low-end CCD camera. For basic observing and recording all you need is a television set for a monitor and a VCR. These two items are much less expensive than a personal computer, and many households have more than one of each. Most people would be willing to expose this equipment to the night air rather than their PC.

For video capture (known colloquially as "frame-grabbing") and image processing, the VCR and videotape can be moved indoors and connected through a video capture device to a PC. This allows the imager to work in a comfortable environment when the time comes to select desirable images from the video. An added bonus to purchasing this device is that you can also capture images from your daily home videos taken with a camcorder.

The biggest advantage of astro-video cameras in lunar and planetary videography is that the simplest telescope can be used. With such fast exposure times and a frame rate of 30 exposures per second, clock-driven telescopes are not absolutely necessary. An observer can use an altazimuth telescope and manually "push" the telescope around to keep the image centered. This is just as applicable to a 60 mm ($2\frac{1}{2}$-inch) refractor as to a 630 mm (25-inch) Dobsonian reflector. No longer are expensive equatorial mountings and high-precision clock drives necessary to take great pictures of the Moon, the Sun, and the bright planets. This makes the astro-video camera highly appealing to an amateur who wants good results on a small budget.

What telescope?

Almost any telescope can be used for imaging the Moon with an astro-video camera, from a 60 mm refractor to the largest observatory instrument. One has even been used on the 60-inch (1.5 m) reflector on Mount Wilson. The telescope of course must be

Figure 12.5. Joe LaVigne's 250 mm (10-inch) *f*/6 Dobsonian reflector used to image the Moon. *Photograph by Joe LaVigne.*

sturdy enough to support the weight of the camera, and some counterbalancing may be necessary. If the telescope was made to hold 50 mm (2-inch) eyepieces, it should be strong enough to hold the camera. However, when one or more Barlow lenses are placed between the focuser and the camera, then increased leverage is brought to bear on the focuser and some deflection may occur. Additional brackets to take up the load can be made if necessary, or a stouter focuser can be purchased. Figure 12.5 illustrates my 250 mm (10-inch) *f*/6 Dobsonian reflector used to image the Moon.

As stated above, the mounting is not crucial to imaging with an astro-video camera. However, the stability of the mounting, be it altazimuth or equatorial, is important. The motion of an altazimuth must be smooth, and it must dampen out vibrations quickly. This matters because you're going to have to reposition the telescope every few minutes to allow the object to drift through the field of view. The longer the telescope vibrates or flexes, the more unusable the images will be. Fortunately, videotape is a lot cheaper than film.

The advantage to users of equatorial mounts is that precise polar alignment is not necessary. If you have manual slow motions or dual-axis motor control, you simply have to tweak the telescope periodically to recenter the image and obtain superb results. If your equatorial mount is precisely aligned or permanently mounted, you obviously have the ideal setup for imaging. This would allow the Moon or a particular feature to stay centered for long periods of time, allowing you to capture many more of those rare moments of good seeing without the need to shift the telescope and perhaps induce some mild form of motion blur.

Image scale

An important factor to consider in any type of imaging is image scale. The chips in these cameras are relatively small. When they are placed at the focal point of the telescope, they do not necessarily image the entire field of view – they may image only the central portion. The image scale depends on the focal length of the telescope. With the chip at the prime focus of a 60 mm ($2\frac{1}{2}$-inch) refractor, the entire Moon may fit onto the chip with room to spare. On a longer focal length instrument such as a 150 mm (6-inch) $f/8$ or 200 mm (8-inch) $f/6$, you may not be able to get the entire Moon in the field at the prime focus. On systems of even longer focal length, only a part of the Moon will fit. So don't expect to zoom into the crater Copernicus with a small telescope without some form of amplification, and don't expect a small crescent Moon when using a 200 mm Schmidt–Cassegrain telescope.

To increase the size of the image, a negative lens, usually a Barlow lens, is placed between telescope and camera. This effectively doubles the focal length of the telescope, and enlarges the image formed. To increase the image scale further, a second Barlow can be "stacked" into the first one, further increasing the focal length of the system. To ensure good results it's best if the Barlow lenses have good anti-reflection paint inside their barrels and have coated lenses. This can help reduce any ghosting and washing-out of the image when more than one Barlow is used. Using one or more Barlow lenses provides a dramatic increase in image scale. It's amazing to see craters looming large on your video monitor.

What video?

There are a wide variety of options available for the monitor and video tape recorder. At the low end a simple television and a VHS tape player are quite suitable. At the upper end a high-resolution black and white monitor and Super VHS recorder will provide the highest visual resolution. If you decide

Figure 12.6. The 13-inch television set and VCR used by Joe LaVigne to receive input from the Astrovid 2000 control box and camera attached to the telescope. *Photograph by Joe LaVigne.*

to use a standard VHS recorder, it's better you use a four-head model. Figure 12.6 shows my 13-inch television set and VHS VCR set up to receive input from the Astrovid 2000 control box and camera attached to the telescope.

Most of the cameras on the market provide between 380 and 600 horizontal lines of resolution. What this means is that the camera is capable of providing more detail than a standard TV or VHS player can display. Even though it's best to use a high-resolution monitor and a high-end VCR, you can do just fine with standard, off-the-shelf equipment. If you use a high-resolution black and white monitor and a Super VHS recorder, both of which are capable of displaying over 400 lines, then your output will be of very high resolution. This of course increases the price of the overall system.

All this is well and good if your intention is to provide a stunning display of the Moon for the public, or perhaps create a high-quality video for future use. However, the fun does not have to end here. Using a video capture device, more commonly known as a frame-grabber, a computer user can capture a single frame of video and manipulate the image even further to enhance details or even reveal detail hidden in the image. There are a number of devices available, ranging from the simple Snappy made by Play, Inc. to higher-end video capture cards which are installed into one of the expansion slots of a PC.

Using the software provided with the capture device, it is possible to save the image file in a variety of sizes and file types. Depending on the monitor, you can save an image as large as your monitor to any common file type, such as JPEG, GIF, or Bitmap, or even higher-layered formats such as TIFF or FITS. With some software you can make all the adjustments right there after capturing the frame: contrast, brightness, cropping, to name a few. If the observer prefers, the images can later be manipulated in modern astronomical image processing programs or the popular Adobe *PhotoShop*. With the capability of frame grabbing and image processing, the user can achieve excellent results that can be printed out or posted to the Internet.

A typical observing run

A few preliminary steps are necessary to get the equipment ready for an observing run. The telescope has to be set up in the usual fashion, and the optics allowed to equalize in temperature with the surrounding air. While the telescope equalizes, the astro-video camera and its peripheral equipment can be set up. An output cable is run to the control unit then to the VCR, or directly to the VCR. The VCR outputs to the monitor, and a power supply is normally hooked up to the control unit as well. Figure 12.7 shows the Astrovid 2000 camera inserted into the focuser and an output cable attached to the control box.

Once you're ready to go, you need to find your target. Luckily the Moon is an easy one because of its brightness and size. If you were trying to center a planet on the small chip, it would be more difficult. Once you see the bright Moon in the monitor, rack the focuser in and out a bit till you get a sense of the finest focus possible. You'll be amazed the first time you bring the Moon into focus. Unlike a visual view through an eyepiece, you'll find you don't need to refocus as often because the frame rate of the camera will tend to even out the seeing. You may also want to adjust the brightness levels on your monitor to reduce glare and bring the majority of the features into view.

If the Moon is over- or underexposed on the monitor, a good first step would be to adjust the exposure control. By going to a slower exposure, say

Figure 12.7. The Astrovid 2000 camera inserted into the focuser and an output cable attached to the control box. *Photograph by Joe LaVigne.*

1/60 second, the image will become brighter; going to a faster exposure of perhaps 1/10 000 second will make the image dimmer. If your camera has contrast controls you may want to increase or decrease the contrast. Various lunar features stand out at certain contrast levels. Maria and ray craters leap out on high contrast settings, while certain lunar features look more "natural" at standard or low contrast levels.

If you're using one or more Barlow lenses to zoom in on the lunar terrain, the available light will decrease and the features will become dimmer, so you can increase the exposure to brighten the image. Another way to increase the brightness would be to increase the gain, but then a certain level of noise is introduced. This is imperceptible at first, but it can turn into "snow" if too much gain is applied.

At this point you should have a decent image of the lunar surface on your monitor, but things may not be perfect. If there is a breeze the image will shake with the tube. If there is any vibration from the ground, for example from people walking about, you might experience slight vibration in the image. At higher magnifications, movement and vibration is magnified even further. Try to isolate the telescope from such nuisances if you can.

Seeing is of course a major factor in the quality of the image, although not as much as in standard photography or CCD imaging. Try to avoid imaging the Moon if it's over the roof of a house, a parking lot, a driveway, or any other source of radiant heat. By letting your telescope thermally equalize you can improve your local seeing, so don't expect your best images within an hour of setting up the telescope. Imaging the Moon when it is much lower than 30° above the horizon is an exercise in futility unless you're just observing or displaying the image to others. The extra atmospheric turbulence introduces a lot of bad seeing that even a video camera running at 30 frames per second can't erase, and your frame-grabs might not be too pretty. If you talk to any experienced lunar and planetary observer, they'll tell you this is true for any form of observation, visual or photographic. In some cases you may not have a choice, and you may want to image a crescent Moon that is close to the horizon. Then you'll just have to be satisfied by whatever you get.

Once you're rolling and have your focus, exposure, contrast, and gain set to your satisfaction, you'll want to start recording. Use good-quality tape, and record in standard play (SP) mode for a higher-quality image.

Image processing

When your recording session is over, you can image-process at your leisure with a computer. The commonly available Snappy video capture device hooks up externally to the parallel port of a PC with the video input coming from the VCR. Video capture cards usually come with software to allow the user to select single frames of video, and some of them allow the user to view the source from an external monitor as well. This is a help if your computer monitor cannot display the video input at its highest resolution.

Before capturing a frame, it helps if you view a few minutes of the tape before the actual capture. This will allow you to pick a favorable patch of seeing on the tape. Once you've found a desirable patch, note where it is on the tape and rewind to just before that section. As you play the tape again and approach the section, hit "Pause." Most four-head VCRs have the ability to advance the tape frame by frame, either by hitting "Pause" repeatedly or by holding it down and letting it advance automatically. This lets you watch each frame and choose the sharpest one available.

Sometimes it's hard to find a single frame that stands out from the rest. If this is so, watch for very small and fine details on the lunar surface. As you advance frame by frame, look for fine lines that suddenly leap out and become sharp. More than likely they will blur out in surrounding frames, and this will enable you to home in on the sharpest image available on that section of tape. Once you've found your frame, command your software to capture it.

Once the frame is captured you can use either the capture software or other image processing software to save and manipulate the file. Some formats save into larger files than others. Bitmap and TIFF files can be rather large, while JPEG and GIF files don't take up as much storage space. This can be a consideration if you plan to take a lot of frame grabs over a period of time, but then again large hard drives are becoming very inexpensive.

Lunar detail tends to be fairly contrasty, and this determines the type of image processing that will be of use to the lunar observer. Techniques that are used in deep-sky and planetary image processing, such as image stacking and unsharp masking, are of limited value when used on lunar images. These techniques work well to increase detail by increasing the contrast of faint and subtle features of nebulae and planets, but with lunar images they tend to make dark features darker, sometimes obscuring subtle details that are hiding in the shadows.

Use the Contrast and Brightness controls of the image processing software to gently manipulate the image. If the image is bright, use the brightness control to tone it down a bit. If the image is a bit dark, an increase in contrast will actually brighten it up and reveal features. Another technique to bring out detail is to use the Sharpen controls. You need to be careful about how much you do this, because it can introduce noise into the image and even ruin it. However, it can be surprising how much an image can be improved by sharpening. Cropping allows the user to trim the image to display a prominent lunar feature, while Zoom is great for adding that extra bit of "focal length" and enlarging those subtle rilles and craterlets. Figure 12.8 is a video image of the crater Clavius, frame-grabbed from video and enhanced using the Zoom, Sharpen, and Contrast controls in Adobe *PhotoShop 4.0*.

A multitude of possibilities

Astro-videography is a very versatile tool for the lunar observer. For casual observing and public events, this medium provides an elegant way to see the myriad features of the Moon. Used with a good lunar atlas, an astro-video can help an observer learn the topography of the Moon in short order. For observers wishing to make a serious study of lunar geography or geology, astro-video opens up a multitude of possibilities, limited only by the imagination of the observer. Considering the cost of CCD cameras and today's high-tech telescopes, astro-video is a modest investment with a big payoff.

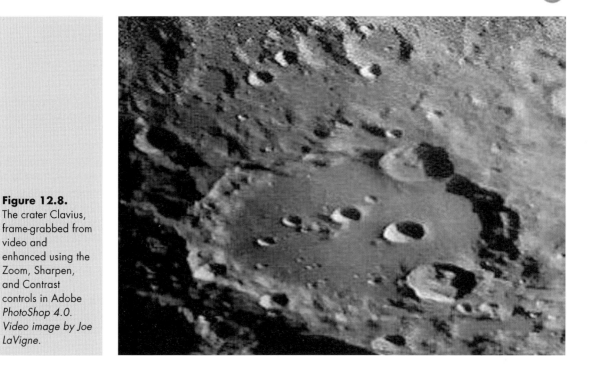

Figure 12.8.
The crater Clavius, frame-grabbed from video and enhanced using the Zoom, Sharpen, and Contrast controls in Adobe PhotoShop 4.0. Video image by Joe LaVigne.

The CD-ROM Lunar Atlas

On the CD-ROM accompanying this book is an atlas of selected lunar surface features, for your reference and enjoyment. Many of the more interesting formations described in Chapters 4 to 7 are illustrated on this CD-ROM. The atlas is the work of an amateur astronomer, Eric Douglass, who uses a video capture technique similar to that described in the last section, but with somewhat different equipment. In compiling the atlas, he employed a Burle TC-650 monochrome video camera with 380 lines of horizontal output, a lux rating of 0.015, and 250 000 pixels arranged on a $\frac{1}{2}$-inch CCD chip. This was attached to his 320 mm ($12\frac{1}{2}$-inch) f/6 Newtonian reflecting telescope using the eyepiece projection technique common in regular astrophotography, with TeleVue Plössls as his chief eyepieces.

The video output from Eric Douglass's set-up is directed to a Videonics Video Equalizer, a type of digital video processor, allowing the observer to adjust contrast and brightness to enhance the raw images. The output from the equalizer next goes to a Super VHS recorder. He recommends using only the very highest quality Super VHS videotapes; he prefers Fuji tapes with a double coating, for their high "luminance" rating. After the images of the lunar features are recorded on the tapes, the recorder is taken inside and hooked up to a Snappy frame-grabber, which allows individual frames to be stored and processed on a PC. The PC is equipped with Fauve Matisse imaging processing software that will detect and subtract, noise, add frames together, or splice images.

The end product of all this hard work is the atlas on the CD-ROM, which contains 102 images of many of the more prominent lunar features. Of particular interest are the two very unusual and contrasty full moon mosaic images, one of which is reproduced as Figure 3.4, that are among the most beautiful and scientifically interesting that I have ever encountered. They were taken using the video system described above, but with a special near-infrared filter, with a 50% transition at a wavelength of 780 nm. The mosaics show excellent contrast between the dark, basalt-flooded mare regions and the bright brecciated lunar highland crust. You will want to refer to these images over and over again when reading the chapters on lunar formations and lunar geology.

Eric Douglass, who finds the Moon both geologically fascinating and aesthetically pleasing to observe, offers this advice: "Learn to *see*, not just look … pay attention to the details … the meaning of the geology is in the details." He recommends using a wide variety of magnifications, from very low power to the highest that your observing conditions will allow, when observing the Moon through your telescope, each magnification revealing different geologic and aesthetic aspects of the lunar surface. "Observing is more than the sum of its parts." I couldn't agree more!

When I was in astronomy graduate school back in the mid-1980s, professional astronomers were getting very excited by the possibilities of what were called charge-coupled devices, or CCDs, a magical application of the still fairly new silicon semi-conductor chip. CCDs quickly began to replace older technologies like hypersensitized photographic emulsions and photomultiplier tubes. But few would have thought that this same advanced technology would soon be readily available to amateur astronomers, much less that it would be affordable to them. Because CCD imaging requires a computer to interpret and process the electronic images formed by the detector, the development of CCD technology would not have been possible without the PC revolution of the 1990s.

There is now a bewildering array of CCD cameras, and related software products available to the amateur astronomer, and it would take a whole book just to survey them. By the same token, CCD imaging is indeed high-tech work, and you will have to spend some time learning the basic operation of whatever CCD you acquire, as well as the ins and outs of whatever software you use with it, before you will be able to start imaging. Because each CCD and software program is different, it is beyond the scope of a book like this to explain in detail the usage of the products. For a general background on CCDs and their use, I would recommend another of the books in this series, *The Art and Science of CCD Astronomy*, edited by David Ratledge.

Instead, it seems to make better sense to briefly sketch out the functioning of one of the more popular CCD cameras on the market today, the Starlight Xpress (the company actually makes several different cameras, typical of CCD suppliers) and give examples of the kinds of lunar image that are obtainable with this device. The following section was contributed by Maurizio Di Sciullo, an amateur astronomer who has achieved considerable proficiency with the Starlight Xpress CCD, despite having worked with this device for little more than a year. When you see some of his lunar images that appear in this chapter, you will be astounded by their detail and beauty, and you will almost certainly want to try your own hand at CCD imaging of the Moon.

Lunar CCD Imaging

Maurizio Di Sciullo

Imaging the Moon with modern CCD equipment can be extremely enjoyable and entertaining. Aside from the entertainment value of CCD imaging, one stands to learn much about lunar geography, as well as the mechanics that created our Solar System. The Moon presents us with many unique possibilities in imaging, and is therefore the natural choice for beginning CCD imagers as their first subject.

The CCD camera is basically a device that collects photons from a source, translates the intensity of photon strikes into a digital value, and converts the numeric data into an analog image that is displayed on a PC. Like everything else, CCDs have their limitations. Their sensitivity to light is not perfect, their spectral responses are rarely flat, and their resolution is dictated to some degree by a complex relationship between pixel size, atmospherics, the subject's luminosity, and the quality of the optics delivering the image to the camera. These obstacles have frustrated many a prospective CCD imager, but fortunately they are minimized when imaging the Moon.

Several factors that haunt deep-space CCD imagers are absent when imaging the Moon. Specifically, as the Moon is essentially sitting in broad daylight, only 363 263 km away from the Earth at perigee, there is no lack of photons. Indeed, the Moon is regularly visible in daylight. This abundance of photons removes the problems associated with long exposure times – at moderate focal ratios, modern CCD cameras can be exposed for as little as 0.03 s or less on the first quarter Moon.

In a "domino effect", this reduction in exposure times means that lunar imagers do not have to rely on their mounting's accuracy, as deep-sky observers do. Further, the shorter exposures also allow the camera to "freeze" exposures, or even to complete an exposure in a fleeting moment of steadiness between periods of atmospheric distortion, thereby increasing resolution. Finally, the Moon's inherent brightness can be used to trade off photon flux for image scale, allowing greater magnifications to be used, if the optics can withstand it. It is this assortment of seemingly trivial factors that makes the Moon an excellent choice as a novice CCD imager's first subject.

Acquiring the images

My first endeavor with a CCD camera was to image the Moon. After very little trial and error, I obtained

amazingly good results. (I have since developed various image-processing techniques, so I shall present as examples only images I acquired in the early days of experimentation.) As a subject, the Moon offers a huge amount of detail, both large and microscopic. The larger features require only the eye to detect, while the fine rilles and craters require the aid of a good telescope. Coupled with a modern CCD camera, high-performance telescopes can realistically resolve features on the Moon down to the size of a football field The main limitations are seeing and optical quality.

The telescope I used to acquire the images in this section was an Excelsior Optics E-258, a 254 mm (10-inch) planetary Newtonian reflector. The E-258 has many of the desired qualities for a CCD imaging telescope – reasonable aperture, long enough focal length (f/8) to provide a generous image scale, and irreproachable optics – to fully exploit those precious moments of excellent seeing that astronomers so covet. I used a Starlight Xpress HX-516 CCD camera for the imaging, as it also provides many of the features that astronomers and imagers alike find desirable. One of the Starlight Xpress CCD cameras, very similar to the one I used, is illustrated in Figure 12.9.

The Moon offers its greatest rewards to CCD imagers who pursue the beauty of the first quarter terminator. Along this great expanse of long, inky shadows, and brilliant, jutting spires, one finds what is easily the most challenging visual dynamic range in the sky. At one extreme, the darkest shadows of the Moon are illuminated only by the feeble light of earthshine, or by the equally feeble light reflected from the illuminated surfaces of nearby mountains. At the other extreme, light-colored mountains and ejecta-splashed terrain are subjected to the blinding intensity of direct sunlight, with no atmosphere to act as a buffer. Further, given the effects of solar angle and the dynamics of reflection and dispersive phenomena, the surface brightness of the Moon near the terminator changes drastically over just a few arc minutes.

This dynamic range, which is easily a trillion to one, is impossible for modern CCD cameras to digitize properly. As a result, prospective lunar CCD imagers must adjust their imaging programs accordingly, so that their frames do not span areas of extreme brightness variation, unless absolutely necessary. Two CCD images I took with the Starlight Xpress HX-516 that appear as illustrations in earlier chapters of this book – Figs 1.1 and 2.11 – illustrate the dynamic range you have to deal with when imaging the lunar landscape. Figure 1.1 shows the bright Apennine Mountains contrasted against the dark of the Imbrium plain, while Figure 2.11 is a close-up of the Apennines that shows their steep cliffs casting very dark shadows almost immediately below their very bright white peaks.

The initial setup for the lunar imaging apparatus was very straightforward. I determined that a focal length of 4000 mm (157 inches) would be appropriate for the initial trials, as this corresponded nicely to the E-258's focal length with a simple Barlow lens included to double the f-ratio. Using a TeleVue 50 mm (2-inch) ×2 Big Barlow, I configured a simple and effective prime focus projection system that

Figure 12.9.
A Starlight Xpress CCD camera.
Courtesy Starlight Xpress.

would allow excellent resolution of all but the finest lunar features. There was no complicated eyepiece projection apparatus, nor was any special attention given to strict image "sampling" calculations which so many deem indispensable for "proper" CCD imaging. You will quickly find that lunar imaging programs are forgiving of all but atmospheric snafus, provided that you employ at the very least a minimum of common sense.

My first images showed that theory and practice were consistent. The exposure times were just 0.05 s, which reduced the corrupting effect of atmospheric distortions to a minimum and also helped to safeguard against blurring resulting from drifts caused by minor errors in the mount's tracking and pointing. Theory and practice also agreed on the optical results that streamed off the monitor – the E-258 was resolving features down to pixel limitations, of approximately 0″.33 per pixel. It was obvious from the initial data that the magnification could have been increased even further, and later tests during

periods of excellent seeing yielded a best resolution of 0″.09, after processing.

Once the CCD camera had delivered the raw data, it was a trivial matter to process it into finished frames. Processing binary image files can be extremely complicated, easily blurring the line between science and art. The best image processors do most of their work by "feel," relying little on any hard and fast rules for numerical manipulations. In the case of the Moon, the novice should strive to avoid complex processing routines, concentrating instead on choosing the best raw frame which will require a minimum of processing. In fact, if I had to choose a single hard and fast rule for image processing, it would be "less is better."

The image processor's most powerful tool is what is referred to as *unsharp masking*. Briefly, this works by taking a weighted average brightness of the area surrounding each pixel in the image, and then increasing the difference between the pixel and that average. This process enhances small-scale image

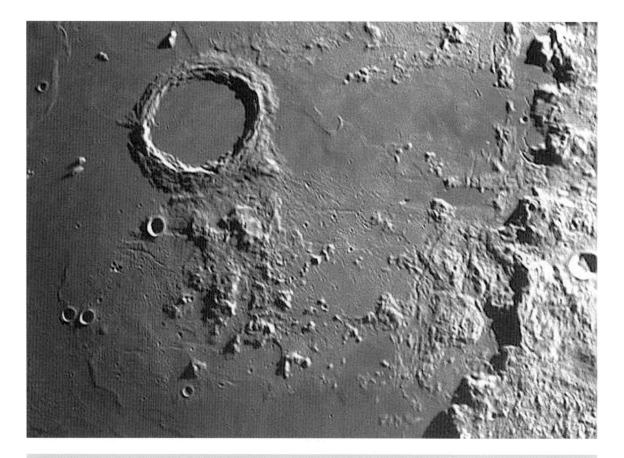

Figure 12.10. A wide-angle view of the Apennines/Archimedes region. Hadley Rille, explored during the Apollo 15 mission, is at the extreme upper right of the frame. *CCD image by Maurizio Di Sciullo.*

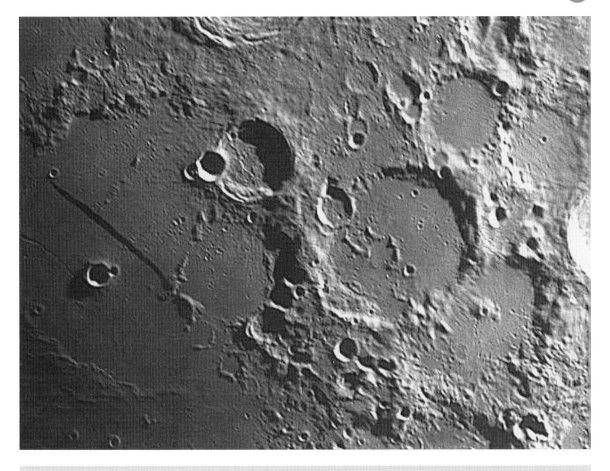

Figure 12.11. The famous Straight Wall and the large craters Purbach and Regiomontanus. *CCD image by Maurizio Di Sciullo.*

structure without disturbing the overall brightness of an image. The degree of enhancement is determined by the chosen radius and strength of the applied mask. The radius of the mask refers to how many pixels from the measured spot are averaged, and the weight refers to how intensely the difference between the point and surrounding area is magnified. A similar effect is accomplished by what are referred to as fast Fourier transforms (FFTs), which serve to enhance the crispness of an image. For lunar processing, beginners will find that simple unsharp masking will serve them well, and should strive to perfect this technique before trying out more complex processing and deconvolution techniques.

By using the image processing software provided with the Starlight Xpress camera, I was easily able to achieve sub-arcsecond resolution from lunar images with a minimum of effort. With the best raw images the typical mask radius was only two pixels, the strength being decided by the terrain captured in

each frame. Using only unsharp masks of small radius, I acquired the four images shown here in my first four CCD imaging sessions.

Figure 12.10 is a wide-angle view of the Apennines/Archimedes region; Hadley Rille, explored during the Apollo 15 mission, appears at the extreme upper right of the frame. Notice the changing brightness gradient from the terminator side of the frame (left) toward the illuminated limb, with the associated saturation on the far side of the Apennines. The famous Straight Wall is illustrated in Fig. 12.11, along with the large craters Purbach and Regiomontanus. Figure 12.12 depicts the Heraclitus/Licetus region, pocked with numerous large, eroded craters. Figure 12.13 is an image of the magnificent crater pair Arzachel and Alphonsus, a fine example of the sub-arcsecond resolution possible without a lot of exotic equipment. Here is further proof that there is no substitute for excellent optics.

To prepare the would-be lunar imager for what to expect in the way of resolution, refer back to

Figure 12.12. The Heraclitus/Licetus region, pocked with numerous large, eroded craters. *CCD image by Maurizio Di Sciullo.*

Table 8.1, which gives the smallest sizes of craters and rilles that can be resolved by telescopes of various apertures. The age-old rule of dividing the number 110 by your telescope's aperture in millimeters (or 4.56 by aperture in inches) is hardly an absolute gauge for a telescope's resolving abilities. Go back to Chapter 2 and take a look at Figure 2.12, a CCD image I obtained of the Alpine Valley and its vicinity. Despite the unfavorable solar angle, the famously elusive thread-like rille on the valley floor is visible, testifying to the excellent optics used for the shot. This rille is not supposed to be within the theoretical resolving power of a 250 mm (10-inch) telescope! Obviously, the formulas are not always right – far more important for the aspiring CCD imager are the quality of the optics and the stability of the atmosphere at the observing site. Given those two fairly straightforward constraints, imaging the Moon on a regular basis becomes a surprisingly simple and uncomplicated affair.

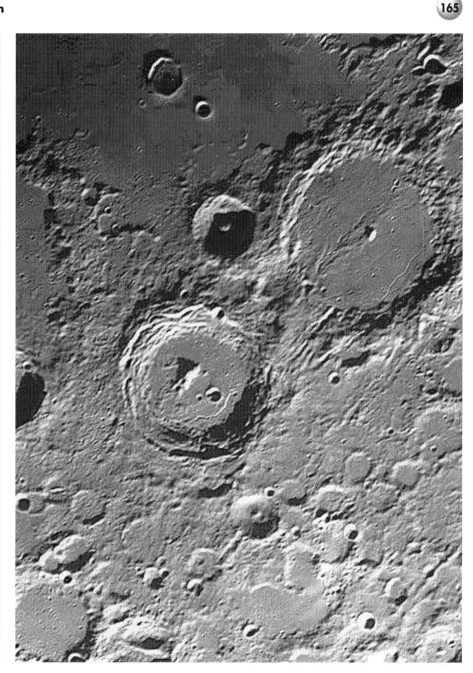

Figure 12.13.
The craters Arzachel
and Alphonsus.
CCD image by
Maurizio Di Sciullo.

Glossary of Lunar Terminology

albedo   A measure of the reflectivity of the Moon's visible surface. The Moon's albedo averages 0.07, which means that its surface reflects, on average, 7% of the light falling on it.

anorthosite   A coarse-grained rock, largely composed of calcium feldspar, common on the Moon.

basalt   A type of fine-grained volcanic rock containing the minerals pyroxene and plagioclase (calcium feldspar). Mare basalts are rich in iron and titanium, while highland basalts are high in aluminum.

breccia   A rock composed of a matrix of larger, angular stony fragments and a finer, binding component.

caldera   A type of volcanic crater formed primarily by a sinking of its floor rather than by the ejection of lava.

central peak   A mountainous landform at or near the center of certain lunar craters, possibly formed by an upwelling of material after an impact event.

cleft   see rille.

co-longitude   see selenographic co-longitude.

concentric crater   A crater consisting of multiple rings, one inside the other, and so looks something like a bulls-eye. Hesiodus and Taruntius are good examples of concentric craters.

crater   A physical depression, of approximately circular outline, on the Moon's surface.

crater cone   The hill, often steep, that a volcanic crater may build about itself.

craterlet   Any lunar crater that is less than approximately 8 km (5 miles) in diameter.

crater pit   A very small craterlet.

dome   A low, rounded elevation on the Moon, possibly of volcanic origin. Some may be as much as 16 km (10 miles) in diameter, but only 15–60 m (50–200 ft) high.

earthshine   The dim illumination of that part of the Moon not lit by the direct light of the Sun. It is most easily seen around the time of the crescent Moon, and is caused by sunlight reflected from the Earth.

ejecta hypothesis   The hypothesis that lunar rays, and certain other phenomena, are formed by solid bodies ejected from a lunar crater during impact.

fault   A fracture of the lunar surface, along which there has been slippage, either vertical or horizontal; a sign of tectonic activity.

feldspar   An alumino-silicate material.

foreshortening   An optical effect that distorts the apparent dimensions of features near the lunar limb. Features near the limb appear crowded together in a radial direction.

gabbro   A coarse crystalline rock, often found in the lunar highlands, containing plagioclase and pyroxene. Anorthositic gabbros contain 65–78% calcium feldspar.

gardening   The process by which the Moon's surface is mixed with deeper layers, mainly as a result of meteoritic bombardment.

ghost crater (ruined crater)   The faint outline that remains of a lunar crater that has been largely erased by some later action, usually lava flooding.

glacis   A gently sloping bank; an old term for the outer slope of a crater's walls.

graben   A sunken area between faults.

highlands   The Moon's lighter-colored regions, which are higher than their surroundings and thus not covered by dark lavas. Most highland features are the rims or central peaks of impact sites.

impact site   A crater or other feature on the Moon formed by the impact of a meteorite, asteroid, or comet with the lunar surface. Craters include everything from huge multi-ringed impact basins like Mare Orientale or Mare Nectaris to minute crater pits.

KREEP   A mineral containing potassium (K), rare earth elements (REE), and phosphorus (P).

libration   The apparent rocking motion of the Moon, either in longitude or latitude. The eccentricity of the Moon's orbit causes libration in longitude. The inclination of its orbit to the celestial equator causes libration in latitude. The fact that we observe the Moon from the surface of an extended and rotating globe (the Earth) itself causes a minor effect, the diurnal libration. In practice, the three types of libration combine to make occasionally visible those regions near the limb that would otherwise be permanently out of view.

limb   The "edge" of the Moon's disk as it appears from the Earth.

lunar transient phenomenon (LTP)   A short-lived and unusual apparent change in the appearance of a lunar feature; also known as a transient lunar phenomenon (TLP).

lunation   The interval between one new moon and the next (or one full moon and the next). It is equal to 29 days, 12 hours, 44 minutes. Also called the synodic month.

mare (*pl.* maria)   A relatively smooth area of the lunar surface, Latin for "sea." Maria are composed of solidified lavas which are darker in hue than the rougher highland areas. It is the dark maria that cause the famous "man in the Moon" effect, seen near the time of full moon.

moonquake   A seismic disturbance in the Moon's interior.

multi-ring(ed) impact basin   A very large impact site, such as Mare Orientale, that displays multiple concentric shock rings, forming a "bulls-eye" pattern of highlands surrounding a central depressed region. Impact sites that are roughly 300 km (200 miles) across or larger were created by meteoroids with sufficient energy to make multiple rings.

occultation   The passage of the Moon in front of a background star, or rarely a planet, as seen from the Earth.

olivine   An iron magnesium silicate mineral commonly found on the Moon.

outgassing   The escape of gas from a lunar formation, probably of volcanic origin. A possible explanation for many LTPs.

oxygen isotopic abundance   In the currently accepted theory of the Solar System's formation, the various chemical elements tended to condense from the primeval solar nebula according to their atomic weights, the heavier ones condensing closer to the Sun, the lighter ones farther away. That is why the planets closest to the Sun tend to contain the heavier elements, and you can tell how far from the Sun a planet (or piece of planet) formed just by looking at its chemical composition. Isotopes are variants of the same chemical element, with different numbers of atomic particles in the atomic nucleus. The ratio of two different isotopes of oxygen is known as the oxygen isotopic abundance, and this quantity is very nearly the same for lunar and terrestrial material.

palus   Latin for "marsh." A lunar area that looked to classical observers like a terrestrial marsh. An example is Palus Putredinis.

phase   The fraction of the Moon's Earth-facing hemisphere illuminated by the Sun.

plagioclase   A calcium feldspar mineral commonly found on the Moon.

plate tectonics   The theory that the Earth's crust consists of huge plates that move away from ridges and either drop downwards or slide laterally where they may collide with other plates, often building mountain ranges in the process. There is little or no evidence of plate tectonics on the Moon.

pyroxene   The ortho-form is an iron magnesium silicate mineral commonly found on the Moon. The clino-form also contains calcium.

rampart   An old term for a crater's walls.

ray system   Bright streaks crossing the lunar surface, radiating from certain craters. They are most apparent around the time of full moon.

regolith   The surface layer of the Moon, composed of a dark, fine, dusty material.

rille (rima, cleft)   A long and narrow, linear or sinuous depression in the lunar surface.

ringed plain   see walled plain.

ruined crater   see ghost crater.

scarp   A cliff or line of cliffs on the Moon (see also wrinkle ridge).

seeing   The quality of the steadiness and sharpness of a telescopic image, as affected by atmospheric and thermal conditions.

selenographic co-longitude   The lunar longitude of the sunrise terminator. Co-longitude is more accurate than phase for indicating the position of the Sun relative to the Moon because it has been corrected for the Moon's librations.

selenography   The study of the location and dimensions of the Moon's physical features (by analogy with "geography"); lunar mapping.

selenology   A now seldom-used term for the study of the Moon's composition and physical nature. Today, "lunar geology" is used instead.

sinus   Latin for "bay," a lava-filled impact site located at the edge of any of the lunar maria. A well known example is Sinus Iridum, on the "shore" of Mare Imbrium.

Sun angle   The angle at which the Sun appears with reference to a feature under observation.

terminator   The great circle on the Moon that is the boundary between day and night. Lunar features are best seen when they are on or near the terminator, as the Sun's low angle creates shadows that emphasize vertical relief.

terraced crater   A crater with inner walls that have terraces – concentric ledges. Terraced craters usually show sunken floors and a well-defined central peak.

walled plain (ringed plain)   A lunar feature consisting of a mare-like floor surrounded by a sharp-crested rim; an old name for a large crater.

wrinkle ridge   An elevated, usually sinuous feature common on the lunar maria, but of uncertain origin. Where they extend to the highlands they are known as scarps.

Further Reading

Alter, D, Lunar atlas, North American Aviation, Downey, CA, 1964; reprinted by Dover, New York, 1968.

Alter, D, Pictorial guide to the Moon, 2nd edn. Thomas Crowell, New York, 1973 [1st edn 1967].

Baldwin, RB, The face of the Moon. University of Chicago Press, Chicago, 1949.

Baldwin, RB, The measure of the Moon. University of Chicago Press, Chicago, 1963.

Barabashov, NP, AA Mikhailov and YN Lipskiy, An atlas of the Moon's far side: The Lunik III reconnaissance. Sky Publishing, Cambridge, MA, 1961 [1st edn 1960].

Benton, JL, A manual for observing the Moon: The ALPO Lunar Selected Areas Program, 4th rev. edn. Association of Lunar and Planetary Observers, San Francisco, 1996.

Bishop, R (ed), Observer's handbook 1999. Royal Astronomical Society of Canada, Toronto, 1998.

Bowker, DE and JK Hughes, Lunar Orbiter photographic atlas of the Moon, NASA SP-206. US Government Printing Office, Washington, DC, 1971.

British Astronomical Association, Guide to observing the Moon. Enslow, Aldershot, UK, 1986.

British Astronomical Association, The handbook of the British Astronomical Association 2000. London, 1999.

Cadogan, PH, The Moon: Our sister planet. Cambridge University Press, Cambridge, 1981.

Cherrington, EH Jr, Exploring the Moon through binoculars and small telescopes. Dover, New York, 1984.

Cook, J (ed) The Hatfield photographic lunar atlas. Springer, London, 1998.

Daly, RA, Origin of the Moon and its topography, Harvard Reprint Series, II, 14, American Philosophical Society, Philadelphia, 1946, pp 104–19.

Dobbins, TA, et al., Introduction to observing and photographing the solar system. Willmann-Bell, Richmond, VA, 1988.

Elger, TG, The Moon: A full description and map of its principal physical features. George Phillip, London, 1895.

Gilbert, GK, The Moon's face: A study of the origin of its surface features. Bulletin of the Philosophical Society of Washington, 12:241–92, 1893.

Goodacre, W, The Moon, Pardy & Son, Bournemouth, 1931.

Graham, FG and JE Westfall, Lunar eclipse handbook, Lunar Press, East Pittsburgh, PA, 1990.

Greeley, R and PH. Schultz (eds), A primer in lunar geology, Comment edn. NASA Ames Research Center, Mountain View, CA, 1974.

Gutschewski, GL, DC Kinsler and EA Whitaker, Atlas and gazetteer of the near side of the Moon, NASA SP-241. US Government Printing Office, Washington, DC, 1971.

Hartmann, WK, Radial structures surrounding lunar basins. I. The Imbrium system. Communications of the Lunar and Planetary Laboratory, 2(24):1–15, 1963.

Hartmann, WK, Radial structures surrounding lunar basins. II. Orientale and other systems; Conclusions. Communications of the Lunar and Planetary Laboratory, 2(36):175–91, 1964.

Hartmann, WK Discovery of multi-ring basins: Gestalt perception in planetary science. In Multi-Ring Basins, Proceedings of the Conference on Multi-Ring Basins: Formation and Evolution, Houston, Texas, 10–12 November 1980, Proceedings of the Lunar and Planetary Science Conference, 12A:79–90, Pergamon, New York, 1981.

Hartmann, WK and GP Kuiper, Concentric structures surrounding lunar basins. Communications of the Lunar and Planetary Laboratory, 1(12):51–66, 1962.

Hartmann, WK, RJ Phillips and GJ Taylor (eds), Origin of the Moon. Lunar and Planetary Institute, Houston, 1986.

Heiken, G, D Vaniman and BM French (eds), Lunar sourcebook: A user's guide to the Moon. Cambridge University Press, Cambridge, 1991.

Hill, H, A portfolio of lunar drawings. Cambridge University Press, New York, 1991.

Jamieson, HD, The lunar dome survey: A progress report. The Strolling Astronomer, 23(11–12): 212–15, 1972.

Jet Propulsion Laboratory, Ranger IX photographs of the Moon: Cameras "A," "B," and "P," NASA SP-112. US Government Printing Office, Washington, DC, 1966.

Kitchin, CR, Telescopes and techniques. Springer, London, 1995.

Kozik, SM, Table and schematic chart of selected lunar objects. Pergamon, New York, 1961.

Kopal, Z, The Moon, our nearest celestial neighbour. Academic Press, New York, 1961.

Kopal, Z, Physics and astronomy of the Moon. Academic Press, New York, 1962.

Kopal, Z, A new photographic atlas of the Moon. Taplinger, New York, 1971.

Kopal, Z and ZK Mikhailov (eds), The Moon, Symposium 14 of the International Astronomical Union, USSR, December 1960. Academic Press, London, 1962.

Kosofsky, LJ and F El-Baz, The Moon as viewed by Lunar Orbiter, NASA SP-200. US Government Printing Office, Washington, DC, 1970.

Kuiper, GP, DWG Arthur, E Moore et al., Photographic lunar atlas, University of Chicago Press, Chicago, 1960.

Kuiper, GP, EA Whitaker, RG Strom et al., Consolidated lunar atlas: Supplements 3 and 4 to the USAF photographic lunar atlas. Lunar and Planetary Laboratory, Tucson, 1967.

Levin, E, DD Viele and LB Eldrenkamp, The Lunar Orbiter missions to the Moon. Scientific American, 218(May):58–78, 1968.

Lewis, HAG, The Times atlas of the Moon. Times Newspapers, London, 1969.

Loewy, MM and MPH Puiseux, Atlas photgraphique de la Lune. Imprimerie Nationale, Paris, 1896–1909.

Lowman, PD, Lunar panorama: A photographic guide to the geology of the Moon. Reinhold Müller, Zurich, 1969.

Lunar and Planetary Laboratory, Communications of the Lunar and Planetary Laboratory, Nos 30, 40, 50, and 70,University of Arizona, Tucson, 1963–66.

Lunar and Planetary Laboratory, Lunar quadrant maps. University of Arizona, Tucson, 1964 (available from Sky Publishing Corp., Cambridge, MA).

MacDonald, TL, The altitudes of lunar craters. Journal of the British Astronomical Association, 39: 314, 1929; 41:172, 1931; 41:228, 1931; 41:288, 1931; 41:367, 1931.

MacDonald, TL, "The Distribution of Lunar Altitudes. Journal of the British Astronomical Association, 41:172, 1931; 41:228, 1931; 41:288, 1931; 41:367, 1931.

Marsden, BG and AGW Cameron (eds), The Earth–Moon system. Plenum Press, New York, 1966.

Maunder, M and PA Moore, The Sun in eclipse. Springer, London, 1997.

Melosh, HJ, Impact cratering: A geological process. Oxford University Press, New York, 1989.

Melosh, HJ, Cratering mechanics: Observational, experimental, and theoretical. Annual Review of Earth and Planetary Science, 8:65–93, 1980.

Middlehurst, BM and GP Kuiper (eds), The Moon, meteorites, and comets, Vol. 4 of The Solar System. University of Chicago Press, Chicago, 1963.

Mobberly, M, Astronomical equipment for amateurs. Springer, London, 1998.

Moore, PA, Guide to the Moon. Norton, New York, 1953.

Moore, PA, The Amateur Astronomer, Lutterworth, London, 1957.

Moore, PA, Lunar Domes. Sky & Telescope, 18(1): 91–5, 1958/9.

Moore, PA, A handbook of practical amateur astronomy. Norton, New York, 1963.

Moore, PA, A survey of the Moon. Norton, New York, 1963.

Moore, PA (ed), Astronomical telescopes and observatories for amateurs. Norton, New York, 1973.

Moore, PA (ed), Practical amateur astronomy. Lutterworth, London, 1973 [1st edn 1963].

Moore, PA, New Guide to the Moon, Norton, New York, 1976

Moore, PA, The modern amateur astronomer. Springer, London, 1995.

Moore, P and PJ Cattermole, The craters of the Moon: An observational approach. Norton, New York, 1967.

Musgrove, RG, Lunar photographs from Apollos 8, 10, and 11, NASA SP-246. US Government Printing Office, Washington, DC, 1971.

Mutch, TA, Geology of the Moon: A stratigraphic view. Princeton University Press, Princeton, 1970.

Nasmyth, J and J Carpenter, The Moon. Scribner & Welford, New York, 1885 [1st edn 1874].

Neison, E, The Moon. Longman, London, 1876.

Ottewell, G, Astronomical calendar 2000, Universal Workshop, Greenville, 1999.

Pickering, WH, The Moon. Doubleday, Page, New York, 1903.

Povenmire, HR, Graze observer's handbook. Privately published, 1979 [1st edn 1975].

Proctor, RA, The Moon. Alfred Brothers, Manchester, 1873.

Quaide, WL and VR Oberbeck, Geology of the Apollo landing sites. Earth-Science Reviews, 5:255–78, 1969.

Ratledge, D, The art and science of CCD astronomy. Springer, London, 1996.

Royal Society, The Moon – A new appraisal from space missions and laboratory analyses. Royal Society, London, 1977.

Rükl, A, Atlas of the Moon. Kalmbach Books, Waukesha, WI, 1992 [1st edn 1990].

Runcorn, SK and HC Urey (eds), The Moon, International Astronomical Union, 1971 Symposium on the Moon, 2nd, University of Newcastle-Upon-Tyne. Reidel, Dordrecht, 1972.

Ryder, G and PD Spudis, Volcanic rocks in the lunar highlands. In: Proceedings of the conference on lunar highlands crust, ed. JJ Papike and RB Merill, Pergamon, New York, 1980, pp 353–75.

Schultz, P, Moon morphology. University of Texas Press, Austin, 1976.

Shoemaker, EM, Exploration of the Moon's surface. American Scientist, 50:99–130, 1962.

Shoemaker, EM, Interpretation of lunar craters. In: Physics and Astronomy of the Moon, ed. Z Kopal. Academic Press, New York, 1962, pp 283–359.

Shoemaker, EM, The geology of the Moon. Scientific American, 211:38–47, 1964.

Sidgwick, JB, Observational astronomy for amateurs. Faber & Faber, London, 1955.

Spudis, PD, The once and future Moon. Smithsonian Institution Press, Washington, DC, 1996.

Stuart-Alexander, D and KA Howard, Lunar maria and circular basins: A review. Icarus, 12:440–56, 1970.

Taylor, SR, Lunar science, a post-Apollo view: Scientific results and insights from the lunar samples. Pergamon Press, New York, 1975.

Thomas, RB, Planetary science: A lunar perspective. Lunar and Planetary Institute, Houston, 1982.

Webb, TW, Celestial objects for common telescopes, 6 edn, Longman, London, 1917, Vol 1, p 77.

Whitaker, EA, GP Kuiper, WK Hartmann et al., Rectified lunar atlas: Supplement 2 to the photographic lunar atlas. University of Arizona Press, Tucson, 1963.

Wilhelms, DE The geologic history of the Moon, US Geological Survey Professional Paper 1348. US Government Printing Office, Washington, DC, 1987.

Wilhelms, DE, To a rocky Moon: A geologist's history of lunar exploration. University of Arizona, Tucson, 1993.

Wilkins, HP, Our Moon. Muller, London, 1954.

Wilkins, HP, Moon maps, with a chart showing the other side of the Moon based upon the Soviet photographs. Faber & Faber, London, 1960.

Wilkins, HP and PA Moore, The Moon: A complete description of the surface of the Moon (containing the 300-inch Wilkins lunar map). Faber & Faber, London, 1961.

Lunar Observing Forms

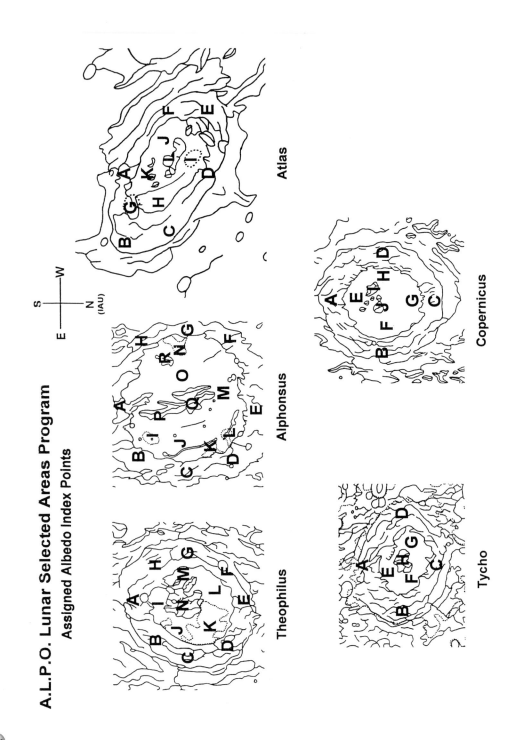

A.L.P.O. Lunar Selected Areas Program
Assigned Albedo Index Points

Atlas

Alphonsus

Copernicus

Theophilus

Tycho

A.L.P.O. Lunar Selected Areas Program
Assigned Albedo Index Points

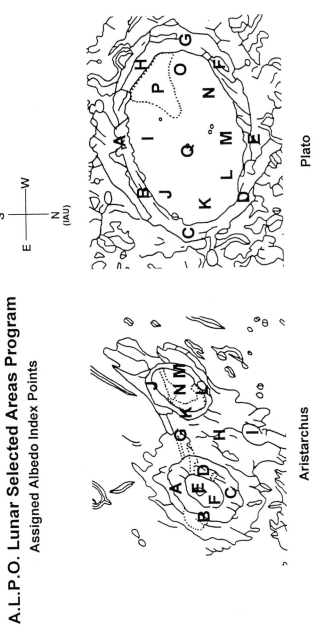

Plato

Aristarchus

South

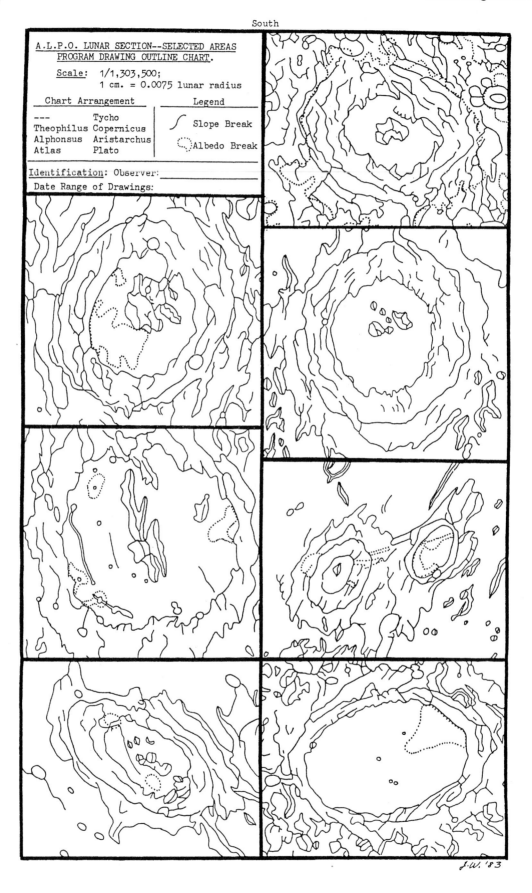

A.L.P.O. LUNAR SECTION--SELECTED AREAS
PROGRAM DRAWING OUTLINE CHART.

Scale: 1/1,303,500;
1 cm. = 0.0075 lunar radius

Chart Arrangement | Legend

Theophilus Copernicus | Slope Break
Alphonsus Aristarchus | Albedo Break
Atlas Plato

Identification: Observer:
Date Range of Drawings:

Tycho

J.W. '83

A.L.P.O. Lunar Section: Selected Areas Program
Albedo and Supporting Data for Lunar Drawings

Lunar Feature Observed : _____
(use *Drawing Outline Chart* for making drawings and attach to this form)

Observer: _____ Observing Station: _____

Mailing Address: _____
  street   city   state   zip

Telescope: _____
  instrument type   aperture (cm.)   focal ratio

Magnification(s): _____X _____X _____X Filter(s): F1 _____ F2 _____

Seeing: _____ [A.L.P.O. Scale = 0.0 (worst) to 10.0 (perfect)]

Transparency: _____ [Faintest star visible to unaided eye]

Date (UT): _____ Time (UT): _____ _____
  year month day   start   end

Colongitude: _____° _____°
  start   end

Albedo Data
(refer to *Albedo Reference Chart* which shows "Assigned Albedo Indices" for feature and attach to this form)

Assigned Albedo Index	Albedo IL	Albedo F1	Albedo F2	Assigned Albedo Index	Albedo IL	Albedo F1	Albedo F2
A				J			
B				K			
C				L			
D				M			
E				N			
F				O			
G				P			
H				Q			
I				R			

NOTES:

Association of Lunar and Planetary Observers: The Lunar Selected Areas Program
Visual Observations of Selected Lunar Features: _____
(indicate chosen feature)

S

Blank for Albedo Indices

N

Drawing Blank

Date (UT):_____ Start Time (UT):_____ End Time (UT):_____

Colongitude (Start):_____ ° Colongitude (End):_____ ° Altitude of Moon:_____ °

Seeing:_____ Transparency·_____ Instrument:_____

Magnification(s):_____X_____X_____X_____X Filter(s): f_1_____ f_2_____ f_3_____ f_4 _____

Observer:_____ Location:_____

Index Point	Albedo: Filter 1	Albedo: Filter 2	Albedo: Filter 3	Albedo: FIlter 4	Index Point	Albedo: Filter 1	Albedo: Filter 2	Albedo: Filter 3	Albedo: Filter 4
A					K				
B					L				
C					M				
D					N				
E					O				
F					P				
G					Q				
H					R				
I					S				
J										

Observational Notes:

A.L.P.O. Lunar Section: Selected Areas Program
Dark Haloed Craters Observing Form

Dark Haloed Crater Observed : _____

(identify by *xi* and *eta* designation and/or *selenographic longitude* and *selenographic latitude*)

Observer: _____ Observing Station: _____

Mailing Address: _____

                       street   city   state   zip

Telescope: _____

                   instrument type   aperture (cm.)   focal ratio

Magnification(s): _____X _____X _____X Filter(s): F1 _____ F2 _____

Seeing: _____ [A.L.P.O. Scale = 0.0 (worst) to 10.0 (perfect)]

Transparency: _____ [Faintest star visible to unaided eye]

Date (UT): _____ Time (UT): _____ _____

        year   month   day   start   end

        Colongitude: _____° _____°

                             start   end

Position of DHC:

xi	*eta*	Selenographic Longitude	Selenographic Latitude	Environs

Lunar Atlas Used as Reference: _____

Dark Haloed Crater (visibility):   Surrounding Dark Halo (visibility):

[ ] definitely visible   [ ] definitely visible

[ ] strongly suspected   [ ] strongly suspected

[ ] vaguely suspected   [ ] vaguely suspected

[ ] not visible   [ ] not visible

[ ] centered   [ ] off center   [ ] circular   [ ] elliptical

                                      [ ] other _____

Relative Intensity (crater) _____   Relative Intensity (halo) _____

Crater Diameter _____ km.   Halo Diameter _____ km.

DRAWING

NOTES:

|_____|

      scale (km.)

A.L.P.O. Lunar Section: Selected Areas Program
Bright and Banded Craters Observing Form

Crater Observed : _____

(identify by name, *xi* and *eta* designation, or *selenographic longitude* and *selenographic latitude*)

Observer: _____ Observing Station: _____

Mailing Address: _____
                        street   city   state   zip

Telescope: _____
                instrument type   aperture (cm.)   focal ratio

Magnification(s): _____X _____X _____X Filter(s): F1 _____ F2 _____
Seeing: _____ [A.L.P.O. Scale = 0.0 (worst) to 10.0 (perfect)]
Transparency: _____ [Faintest star visible to unaided eye]

Date (UT): _____ Time (UT): _____ _____
            year   month   day   start   end

        Colongitude: _____° _____°
                        start   end
Position of Crater:

xi	*eta*	Selen. Long.	Selen. Lat.	Environs

Lunar Atlas Used as Reference: _____

DRAWING

Show detailed morphology, position, orientation, and other characteristics of the crater, including any bands that are definite or suspected, in the drawing blank below. Use the *Albedo and Supporting Data* form for albedo estimates of assigned indices for the crater and for any bands observed (attach to this form). Indicate correct direction of N (IAU) on the drawing.

scale

|------------------------------|

DESCRIPTIVE NOTES:

Index

Files & Lunar Atlas on the CD-ROM

File Name	Resource
Bibliography.pdf	A Selected Lunar Bibliography
Forms.pdf	Lunar Observing Forms
Index.pdf	Files & Lunar Atlas on the CD-ROM
Resources.pdf	Resources for the Lunar Observer
Acrobat\16Bit\AR16E30.EXE	Acrobat Reader (16-bit for Windows 3.1)
Acrobat\32Bit\rs405eng(1).exe	Acrobat Reader (32-bit for Windows 95/98)
Atlas\... .tif	101 lunar images in .tif format (see below)

Lunar Atlas

File names have been chosen to indicate a major or well-known lunar feature that appears in the image, which will, of course, contain other features nearby the feature identified by the file name. In most cases, the reader will be able to easily locate the description in the text of the imaged region simply by looking up the filename in the "CD-ROM Lunar Atlas Cross-Reference". The "Notes" column contains additional information, where necessary, to help the user of the video images locate the precise region of the Moon that appears in the image, so that the text description may be referenced. A * indicates that the specific feature is not separately described in the text.

File Name	Feature Name	Text page no.	Notes
Abulfeda.tif	Abulfeda	57	
Albategnius.tif	Albategnius	55–56	
Alphonsus.tif	Alphonsus	94–95, 127, 130, 163, 165	
AlpineValley.tif	Alpine Valley	24, 50, 164	
AltaiMts.tif	Altai Mountains	20, 57–58, 66	
AriadaeusRille.tif	Ariadaeus Rille	52	
Aristillus.tif	Aristillus	51	
Aristoteles.tif	Aristoteles	49, 50	
Arzachel.tif	Arzachel	25, 95, 163, 265	
Atlas.tif	Atlas	43, 127	
Biela.tif	Biela	—	
Cassini.tif	Cassini	50–51, 79	
Clavius.tif	Clavius	17, 90–92, 103, 117	
Cleomedes.tif	Cleomedes	41, 118	
Copernicus-1.tif Copernicus-2.tif Copernicus-3.tif Copernicus-4.tif Copernicus-5.tif Copernicus-6.tif	Copernicus	18, 25, 35–36, 72–73, 84, 127, 130, 132	Six different views of the crater Copernicus, imaged at different Sun angles, to show different details in the crater formation.
CopernicusRays.tif	Copernicus	18, 25, 35–36, 72–73, 84, 127, 130, 132	The ray system of the crater Copernicus.
Deslandres.tif	Deslandres	—	Region near Hell* and Deslandres* (SW Quadrant).
Eratosthenes.tif	Eratosthenes	24, 35, 72, 151	
Fecunitatis-1.tif	Fecunitatis	19, 24, 29, 38, 47, 60–61, 70, 132–133	Ridge between Mare Fecunditatis and Messier craters.
FullMoon-1.tif FullMoon-2.tif	Full Moon	—	Two mosaics of the Full Moon imaged in the near-Infrared.
Gay-lussac.tif	Gay-Lussac	—	Rille emanating from the crater Gay Lussac*, near Copernicus (NW Quadrant).
Goldschmidt.tif	Goldschmidt	84	
GruithuisenMts.tif	Gruithuisen Mountains	—	Mountains near the crater Gruithuisen* (NW Quadrant).
HadleyRille.tif	Hadley Rille	22, 50, 163	
Hainzel.tif	Hainzel	101	
HeisRidge.tif	Heis Ridge	—	A Ridge near the craters Heis* and Caroline Herschel* (NW Quadrant).
Heraclitus.tif	Heraclitus	163–164	

Hercules.tif	Hercules	43	
Herschel.tif	Herschel	96	
Hippalus-1.tif	Hippalus	100	Two views of the Hippalus rille system, imaged under
Hippalus-2.tif			different Sun angles.
Hipparchus.tif	Hipparchus	55	
Hortensius.tif	Hortensius	84	The Hortensius domes.
Hyginus-1.tif	Hyginus Rille	22, 52	Two views of the Hyginus rille system, imaged under
Hyginus-2.tif			different Sun angles.
Imbrium-1.tif	Imbrium	2, 19, 21, 23–24, 29, 36, 38, 50–51, 73–75, 77–78	Apennine Bench, Montes Archimedes, Palus Putredinis.
Janssen1.tif	Janssen	—	Area near Janssen* (SE Quadrant).
KiesPi.tif	Kies π	26, 100	
Klaproth-1.tif	Klaproth	102	Two views of the crater Klaproth, under different Sun angles.
Klaproth-2.tif			
Langrenus-1.tif	Langrenus	60–61	Two views of the crater Langrenus, imaged under different
Langrenus-2.tif			Sun angles.
Lansberg.tif	Lansberg	99	
Longomontanus.tif	Longomontanus	89, 90	
Maginus.tif	Maginus	13, 92	
Manilius-1.tif	Manilius	53	Two views of the crater Manilius, imaged under different
Manilius-2.tif			Sun angles.
MareCrisium.tif	Mare Crisium	9, 19, 29, 38–41, 44, 47, 55, 132	
MariusDomes.tif	Marius Domes	26, 81	
Maurolycus.tif	Maurolycus	65–66	
Messier.tif	Messier	59–60, 132	Messier & Messier A.
Metius.tif	Metius	68	
Milichius.tif	Milichius	74	Also shows the Milichius dome field.
Moretus-1.tif	Moretus	103	Two views of the crater Moretus, imaged under different
Moretus-2.tif			Sun angles.
MtPiton.tif	Mt Piton	23, 75	
Neander.tif	Neander	—	Area near crater Neander* (SE Quadrant).
Nectaris-1.tif	Nectaris	18–20, 24–25, 38, 58, 60, 63 66–68	Three views of Mare Nectaris, showing different parts of the basin.
Nectaris-2.tif			
Nectaris-3.tif			
Orontius.tif	Orontius	92	
Petavius.tif	Petavius	61–62	
Pico.tif	Piccolimini	58, 66	
Plato.tif	Plato	23, 50, 77–78, 109, 119, 127	
Posidonius.tif	Posidonius	45–46, 49	
Ptolemaeus-1.tif	Ptolemaeus	17, 94, 96	Two views of the crater Ptolemaeus, imaged under different
Ptolemaeus-2.tif			Sun angles.
Purbach.tif	Purbach	93, 163	
Ramsden.tif	Ramsden	99	A view of the Ramsden rille system.
RheitaValley-1.tif	Rheita	68	Two views of the Rheita Valley, imaged under different Sun
RheitaValley-2.tif			angles.
RiphaenMts.tif	Riphaen Mountains	25, 99	
Sacrobosco.tif	Sacrobosco	—	Region near the crater Sacrobosco* (SE Quadrant).
Scheiner.tif	Scheiner	87, 91, 103	
Schiller.tif	Schiller	—	Region near the crater Schiller* SW Quadrant).
SchroetersValley.tif	Schröter's Valley	19, 22, 31, 59, 81–83, 109	
SinusIridum.tif	Sinus Iridum	20–21, 24, 77–79	
Stevinus.tif	Stevinus	64	
Stoefler.tif	Stöfler	65	
StraightWall-1.tif	Straight Wall	93–94, 163	Two views of the Straight Wall, imaged under different Sun
StraightWall-2.tif			angles.
SWlimb.tif	South-western limb	—	Area near southwestern limb that includes Schiller, Segner, Zucchius.*
Theophilus-1.tif	Theophilus	58, 127, 153	Two views of the crater Theophilus, imaged under different
Theophilus-2.tif			Sun angles.
Tycho-1.tif	Tycho	13, 19, 36, 87, 89, 92, 127, 132	Five views of the crater Tycho, imaged under different Sun
Tycho-2.tif			angles.
Tycho-3.tif			
Tycho-4.tif			
Tycho-5.tif			
Vlacq.tif	Vlacq	69, 70	
Walter.tif	Walter	92–93	
Wbond.tif	William Bond	—	Area near crater William Bond* (NE Quadrant).
Weinek.tif	Weinek	—	Area near crater Weinek* (SE Quadrant).
Werner.tif	Werner	57	
Wilhelm.tif	Wilhelm	—	Area near crater Wilhelm* (SW Quadrant).

This CD-ROM contains supplementary material for

Observing the Moon
by Peter Wlasuk

Begin by opening the Index file, **Index.pdf**

If your system already has *Acrobat Reader* installed, you can usually see this simply by clicking (or double-clicking, according to your system) on the file name.

Alternatively, open your *Acrobat Reader*, click on FILE, OPEN, then type the file location and address (such as **D:\ Index**) before pressing ENTER.

If you do NOT have Acrobat Reader on your system, this CD-ROM includes all you need to install it on your IBM-compatible PC, running under Windows 95/98 or Windows 3.x

Two versions of *Acrobat Reader* are included on this CD-ROM. They are located in the **Acrobat** directory. One is a 16-bit version for Windows 3.1, the other is a 32-bit version for Windows 95/98. The reader takes up about 5MB of disk space (Windows 3.1), or 8MB (Windows 95/98).

Windows 95/98:

Make sure all Windows applications are shut down. Use Windows Explorer to explore the sub-folder on the CD-ROM ...**Acrobat\32Bit**. Double-click on the file **rs405eng(1).exe** to begin installation. Follow the on-screen instructions.

Windows 3.1:

Make sure all Windows applications are shut down. Use File Manager to open the CD-ROM subdirectory ...**Acrobat\16Bit**. Double-click on the file **AR16E30.EXE** to begin installation. Follow the on-screen instructions.